Time Series Forecasting using Machine Learning

Tsung-wu Ho

Time Series Forecasting using Machine Learning

Case Studies with R and iForecast

Springer

Tsung-wu Ho
National Taiwan Normal University
Taipei, Taiwan

ISBN 978-3-031-97945-3 ISBN 978-3-031-97946-0 (eBook)
https://doi.org/10.1007/978-3-031-97946-0

This Springer imprint is published by the registered company Springer Nature Switzerland AG
The registered company address is: Gewerbestrasse 11, 6330 Cham, Switzerland

Preface

Machine learning methods are the subject of studies by predictive practitioners but from the perspective of R, the use of machine learning for time series forecasting is a relatively new field. One of the demands, as policy expects, is the need to forecast the real future, and not the back-testing future or a pseudo future that uses cross-validation/training data. While building the model, ACF-type lags may also have to be trained, instead of PACF.

To this end, an R package iForecast is developed that features many machine learning methods, such as SVM, random forest, GBM, elastic net, xgboost and the AutoML routines, and flexible data inputs and lag specifications. In terms of applied econometrics, this text uses a wrapper to estimate VAR modelling using machine learning methods.

This demonstrates the richness and versatility of modern machine learning applications in terms of time series forecasting. It is an introduction to time series forecasting using R, but it is not an introductory statistics/econometrics textbook. Readers should have a basic understanding of matrix algebra and data tables, and we also assumed that readers are familiar with serial dependency concepts, such as ACF/PACF and the implications of serial correlation. This book is intended as an instrumental text for a time series forecasting course, with an emphasis on the use of machine learning methods to forecast a time series future. It will also be of use to economists and econometricians who wish to use R to conduct time series forecasting: specifically, economic time series.

This book does not maintain a commitment to theoretical integrity. It focuses on the practices of forecasting. However, simple theoretical explanations are not omitted so readers who have taken an introductory econometric time series course will find the text beneficial. This course need not necessarily make great use of statistical theory.

Chapter 1 describes the R basics for time series objects, including online data retrieval and the manipulation of time series objects. Internet data retrieval uses several wrappers in the R package JFE, which retrieves data from Fed, Fama French asset pricing data and the Bank of International Settlement.

Chapter 2 explains the ARIMA-based baseline models, covering non-linear models such as SETAR, LSTAR and GAMS.

Chapter 3 explains the use of iForecast to determine ML methods to estimate inputs, which include SVM, random forest, GBM, elastic net and the Auto machine learning routine. Multistep and one-step forecasts are generated, and these are compared with those of Chap. 2.

Chapter 4 discusses the VAR model. A conventional VAR is estimated using Least Squares, but this text uses a wrapper and explains the use of machine learning methods to estimate VAR and generate forecasts. This chapter also includes a section that explains discrete-choice dependent variables.

Chapter 5 details the results for three forecasting case studies: unemployment rate, market index returns and volatility. Unemployment rate forecasting uses the study by Montgomery et al. (1998) and index returns use the study by Chen et al. (1986). The case studies use a rolling windows framework to implement model selection. In practice, the chosen model is used to forecast the real future. The third case study is not detailed in Sect. 5.3 because it uses high-frequency data so its illustration is computationally burdensome. The S&P 500 is used as an example to show the method by which a dataset is compiled for volatility forecasting, which includes daily returns, volume, momentum, the range-based volatility of Parkinson (1980) and the (average) true range that is used by finance practitioners. Interested readers might modify the codes that are used in this book to build their own models.

Chapter-by-Chapter R codes and data can be downloaded from the author's github below:

https://github.com/tsungwu/TSFviaML

I wish to end this preface by thanking my family, friends, students and others who have helped make me what I am today and who by doing so, have contributed to this book.

Taipei, Taiwan Tsung-wu Ho

Competing Interests The author has no competing interests to declare that are relevant to the content of this manuscript.

Contents

Chapter 1
Time Series Basics in R

1.1 Time Series Objects

Time series data has a time stamp and a tabular number. The frame of any data table is at least two-dimensional: rows and columns. On an Excel spreadsheet, columns are horizontally labeled as A, B…, and rows are vertically labeled as 1, 2, 3…, in a series of positive integers. The horizontal columns are variable columns, and the row labels are the row IDs of the variables. Figure 1.1a shows a US real GDP per capita data[1] table. Column ID is a positive integer series, Column 1 Date is the date and the data for the quarterly frequency is identified by the date; Column 2 is GDP and is the data content. Figure 1.1a is not yet a time series, because Date is not an ID label, so date is data.

In Fig. 1.1b, Row ID is replaced with a date and then a time series feature is specified, such as the time zone GMT (Greenwich Mean Time). This data is a standard time series. If Row ID is a positive integer in the computer, it is cross-section data. The cross-section could imply a country code, an industry number, a listed company code, or an individual.

Time series analysis usually requires a lag structure for seasonal modeling and serial dependency, so a time series data structure is important. When a time series data is established, a model can be constructed and forecasting evaluations can be made.

The first step for a time series analysis is collecting the data. This text uses open data from the Internet. This section describes several wrappers that retrieve online time series data.

[1] Source: https://fred.stlouisfed.org/series/A939RX0Q048SBEA

T.-w. Ho, *Time Series Forecasting using Machine Learning*,
https://doi.org/10.1007/978-3-031-97946-0_1

Fig. 1.1 US real GDP per capita. In (**a**), the row names are sequential integers and the Date column is a timestamps data; in (**b**), the row names are replaced by timestamps so it becomes a time series object

```
          Date   GDP
1   1947-01-01 15248
2   1947-04-01 15139
3   1947-07-01 15039
4   1947-10-01 15204
5   1948-01-01 15371
6   1948-04-01 15563
           ...   ...
306 2023-04-01 67295
307 2023-07-01 67916
308 2023-10-01 68351
309 2024-01-01 68549
310 2024-04-01 68977
311 2024-07-01 69399
```

```
GMT
             GDP
2021-07-01 64877
2021-10-01 65986
2022-01-01 65779
2022-04-01 65760
2022-07-01 66115
2022-10-01 66575
2023-01-01 66967
2023-04-01 67295
2023-07-01 67916
2023-10-01 68351
2024-01-01 68549
2024-04-01 68977
2024-07-01 69399
```

```
dat0=read.csv("data/GDP.csv")
timestamp =as.Date(dat0[,1])
```

To construct the `timeSeries` object:

```
dat=timeSeries::timeSeries(dat0[,-1],dateID)
```

or for a `xts` object:

```
dat=xts::xts(dat0[,-1],timestamp)
```

For a `zoo` object:

```
dat=zoo::zoo(as.matrix(dat0[,-1]), timestamp)
```

A time series data can also be constructed by pre-embedding `rownames` with a `timestamp` to allow the transformation:

```
dat=as.matrix(dat0[,-1])
rownames(dat)=as.character(timestamp)
dat=timeSeries::timeSeries(dat)
```

However, this method cannot be used for xts nor zoo, so `timeSeries` is preferable to `xts` and `zoo`. Figure 1.2 plots these four time series plots, where (a) is manipulated by package `timeSeries`, (b) is framed by `xts`, (c) is the graph that is structured by `zoo`, and (d) is R Base `ts`.

These three package-based methods are used for high-frequency or irregular time series, such as, week, day, and intraday (second/minute/hour). `ts()` is not applicable. To illustrate this, the Dow Jones Index is read:

```
dat0=read.csv("data/DJI.csv")
```

Figure 1.3 shows the data table. The first column is the timestamp.

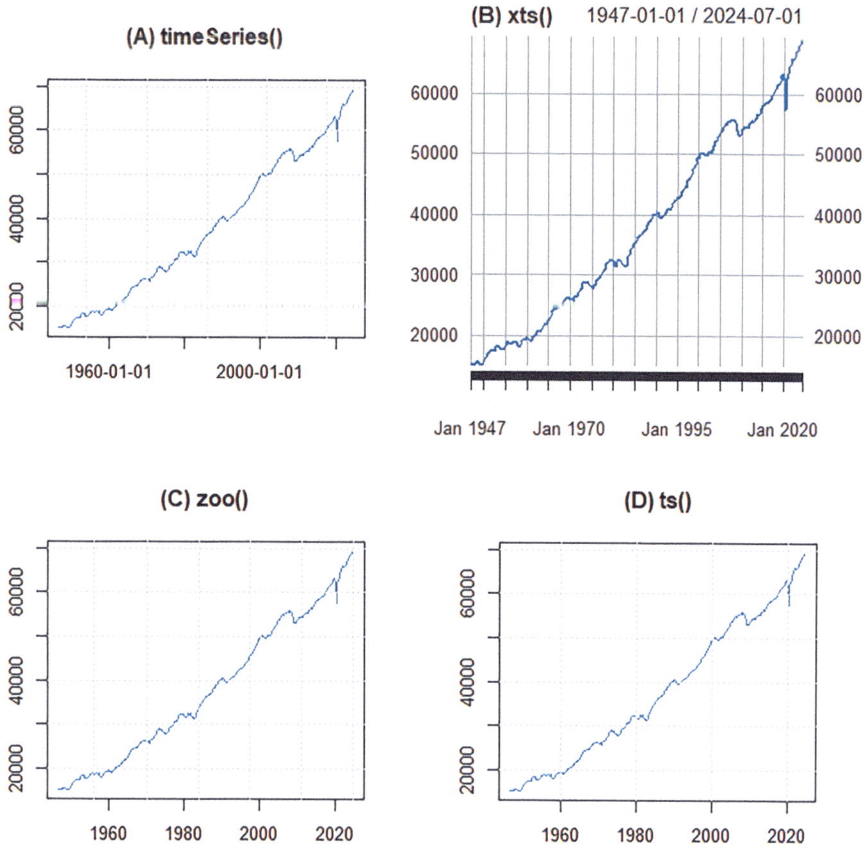

Fig. 1.2 Quarterly data that is manipulated using four time series formats: (**a**) is manipulated by package `timeSeries`, (**b**) is framed by `xts`, (**c**) is structured by `zoo`, and (**d**) is R Base `ts`

The same procedure is then used to transform this into time series object as:

```
timestamp=as.Date(dat0[,1])
timeSeries(dat0[,-1],timestamp)
xts(dat0[,-1],timestamp)
zoo(as.matrix(dat0[,-1]),timestamp)
```

Figure 1.4 shows these plots.

1.2 Accessing Open Data

Throughout this text, open data that is downloaded from the Internet is used. This section describes methods for economic and financial time series.

date	DJI.Open	DJI.High	DJI.Low	DJI.Close	DJI.Volume	DJI.Adjusted
2007-01-03	12459.54	12580.35	12404.82	12474.52	327200000	12474.52
2007-01-04	12473.16	12510.41	12403.86	12480.69	259060000	12480.69
2007-01-05	12480.05	12480.13	12365.41	12398.01	235220000	12398.01
2007-01-08	12392.01	12445.92	12337.37	12423.49	223500000	12423.49
2007-01-09	12424.77	12466.43	12369.17	12416.6	225190000	12416.6
2007-01-10	12417	12451.61	12355.63	12442.16	226570000	12442.16
...	...	...	...	...	...	...
2025-01-16	43290.25	43294.81	43081.75	43153.13	504270000	43153.13
2025-01-17	43312.55	43653.25	43312.55	43487.83	554870000	43487.83
2025-01-21	43528.65	44050.13	43528.65	44025.81	592210000	44025.81
2025-01-22	44178.06	44208.34	44042.11	44156.73	588430000	44156.73
2025-01-23	44113.55	44565.26	44113.55	44565.07	444180000	44565.07
2025-01-24	44533.75	44545.52	44332.22	44424.25	532230000	44424.25

Fig. 1.3 Daily time series data table for the Dow Jones Index, 2007/1/3–2025/1/24, which is retrieved using the `getSymbols()` for package `quantmod`

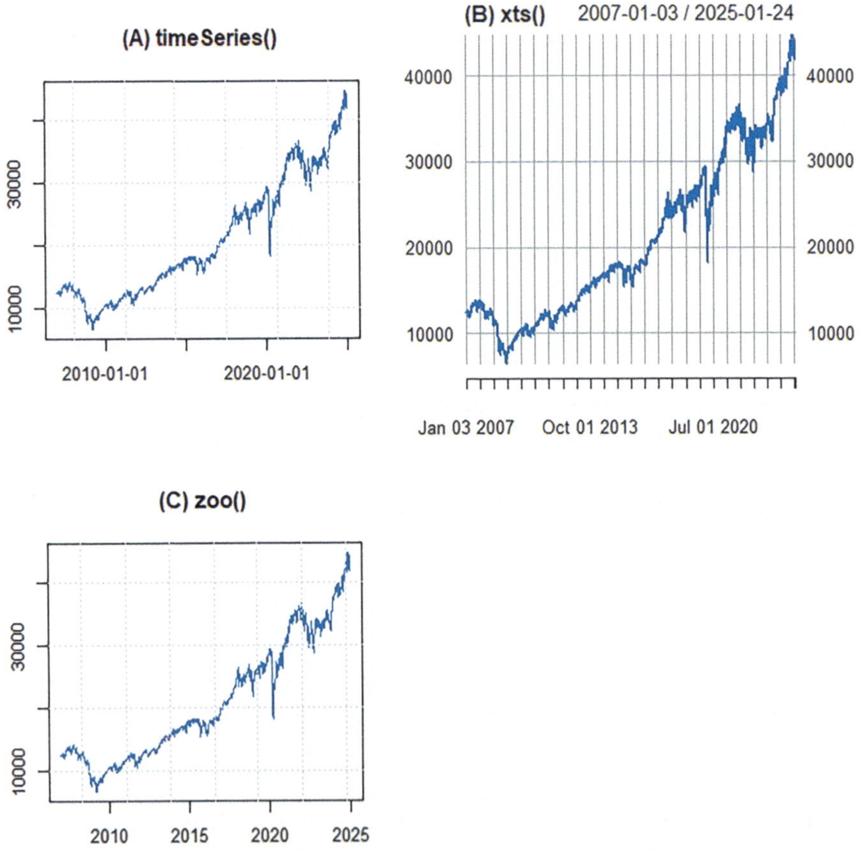

Fig. 1.4 Daily closing price for the Dow Jones Index, manipulated using three time series formats: (**a**) is manipulated by package `timeSeries`, (**b**) is framed by `xts`, and (**c**) is structured by `zoo`

```
> EURUSD
            EUR.USD
2024-08-03 1.09110
2024-08-04 1.09112
2024-08-05 1.09477
2024-08-06 1.09320
2024-08-07 1.09208
2024-08-08 1.09219
2024-08-09 1.09193
2024-08-10 1.09155
2024-08-11 1.09154
2024-08-12 1.09248
        ...
2025-01-19 1.02720
2025-01-20 1.03482
2025-01-21 1.03908
2025-01-22 1.04204
```

Fig. 1.5 EUR.USD gives the US dollar-EURO rate; data source Oanda only provides historical data for the past 180 days

1.2.1 `quantmod::getSymbols` *Retrieves Financial-Economic Data*

The R package `quantmod` uses `getSymbols()` to retrieve data from Yahoo Finance (finance.yahoo.com), which downloads pricing information for equity, commodities, and foreign exchanges.

Data that is retrieved from the `quantmod` is in `xts` format by default. To retrieve many market indices, an object with the name that appears on Yahoo is used. To download the Dow Jones Industrial Index (DJI), the S&P500 (GSPC), and the Nikkei (N225), the system uses:

```
quantmod::getSymbols(c("^DJI","^GSPC", "^N225"))
```

to download the three market indices. Each index has the same content as Fig. 1.3, and the downloaded data is stored in the global environment using the same name as an input symbol, such as `DJI`.

The wrapper `getFX` retrieves US dollar-EURO rate as:

```
quantmod::getFX("EUR/USD", src='oanda')
```

EURUSD appears on the screen as (Fig. 1.5):

The wrapper `getMetals` retrieves Metals, and the most useful feature of `getMetals` is its ability to specify the Metals in terms of 3-character symbols or by name: XAU (gold), XAG (silver), XPD (palladium), or XPT (platinum).[2]

[2] Users should visit https://www.oanda.com/ for details on specific pricing issues.

```
           XAU.USD XAG.USD
2024-08-03 2442.46 28.5625
2024-08-04 2442.37 28.5598
2024-08-05 2416.84 27.6214
2024-08-06 2400.52 27.0654
2024-08-07 2391.32 26.9236
2024-08-08 2407.64 27.1326
2024-08-09 2426.92 27.5078
2024-08-10 2431.25 27.4535
2024-08-11 2431.12 27.4484
2024-08-12 2449.04 27.7658
   ...
2025-01-19 2702.94 30.3510
2025-01-20 2705.75 30.3892
2025-01-21 2730.39 30.6366
2025-01-22 2754.18 30.8266
2025-01-23 2751.65 30.5088
2025-01-24 2773.25 30.7407
2025-01-25 2770.88 30.5985
2025-01-26 2770.76 30.5972
2025-01-27 2752.41 30.2900
2025-01-28 2749.76 30.2148
```

Fig. 1.6 Using `quantmod::getMetals(c("XAU", "silver")))` to retrieve Gold(XAUUSD) and Silver(XAGUSD)

`quantmod::getMetals(c("XAU", "silver")))` retrieves gold and silver market prices in US dollars, as shown in Fig. 1.6.

Similarly, data source Oanda only provides historical data for the last 180 days.

`quantmod::getSymbols` is used and `src='FRED'` is specified to retrieve economic data from FRED:

```
quantmod::getSymbols('CPIAUCNS', src='FRED'),
```

downloads the US Consumer Price Index (CPI).

All `quantmod` wrappers return `xts` format by default and `quantmod::getSymbols` does not require that a left-hand side label is specified. Namely:

```
y=quantmod::getSymbols('CPIAUCNS', src='FRED')
```

the left-hand-side `'y='` is the wrong command for `getSymbols`.

1.2.2 *`JFE::getFed`* *Retrieves Economic Data*

FRED(Federal Reserve Economic Data, fred.stlouisfed.org) hosts many economic time series. To illustrate, the US CPI is downloaded in terms of:

```
cpi_m=JFE::getFed("CPIAUCNS")
timeSeries::as.timeSeries(cpi_m$data)
xts::as.xts(cpi_m$data,
            order.by=as.Date(rownames(cpi_m$data)))
```

Unlike `getSymbols`, the output format for the `getFed` series only embeds the row names with a timestamp and there is no specific default style, so this can be freely transformed later, which is a convenient design.

Although `getFed("CPIAUCNS")` retrieves the same data as `getSymbols('CPIAUCNS', src='FRED')`, there are at least two differences:

Firstly, `JFE::getFed()` can be specified with or without the left-hand side variable.

Secondly, in FRED, most monthly data has a quarterly and annual frequency. The quarterly frequency for the same series is downloaded by specifying `JFE::getFed ("CPIAUCNS",freq="Quarterly")` or `cpi_a=JFE::getFed("CPIAUCNS",freq="Annual")`, which is not supported by `getSymbols`.

For foreign exchange rates, `quantmod::getFX` and `JFE::getFed` are complementary. The exchange rates for the FRED omits the latest 20 days so `JFE::getFed` does not include the most recent observations. `quantmod::getFX` downloads data from `Oanda`, which only gives the most recent 180-day observations so these are stacked them together using `rbind()`:

```
f1=JFE::getFed("DEXUSEU",freq="Daily")
f1=xts::as.xts(f1,
               order.by=as.Date(rownames(f1)))
```

Proper rows are selected from `f1` and stack `EURUSD` (data shown in Fig. 1.5). Exchange rates are worldwide so two series may have a different digital appearance.

When exchange rate data is retrieved, the symbols must be defined properly. The standard is to express foreign exchange rates against the US dollar as the 1-dollar denominator. The exceptions are UK sterling and Euro.

1.2.3 `JFE::getEER` Retrieves BIS Effective Exchange Rates

Firstly, we show how BIS effective exchange rates are retrieved. Using the JFE wrapper `getEER()` eases loading.

The Bank of International Settlement (BIS, https://www.bis.com) contains nominal and real effective exchange rates. Since 2024, BIS has released its data API, which has a friendly interactive interface for three programming languages (Matlab/Python/R). This automatically generates a URL for the data. BIS contains many time series data sequences for global markets. The guide for exchange rates is detailed below:

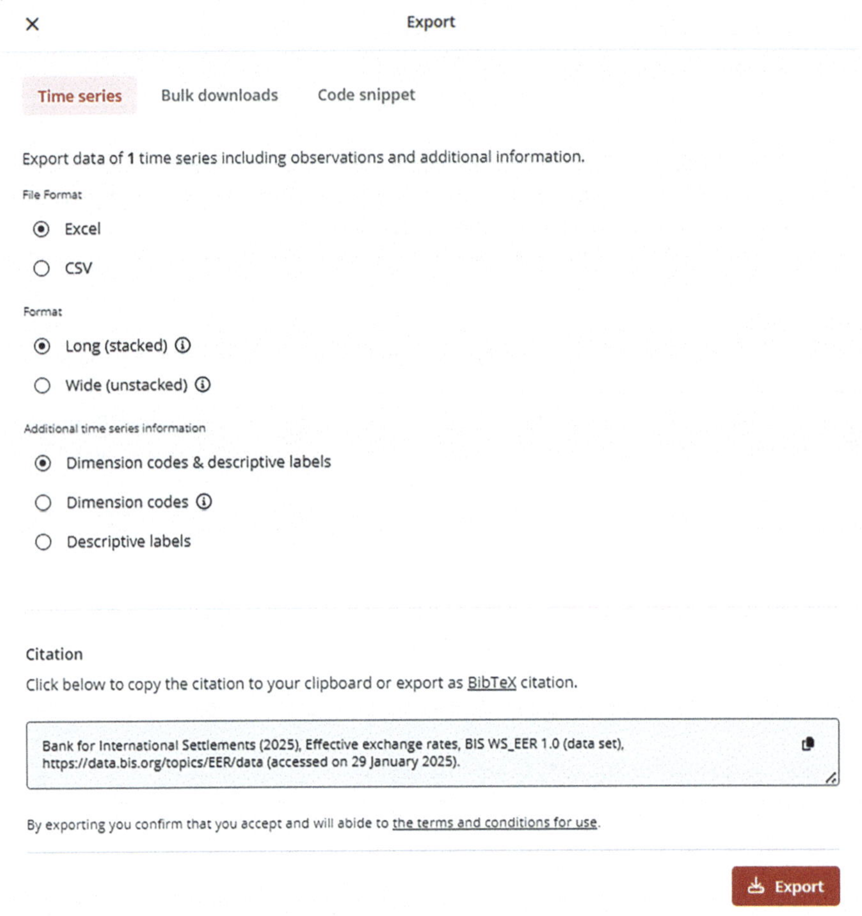

Fig. 1.7 The pop-up **Export** window on the webpage for BIS Statistics

Step 1. Access https://www.bis.org/, go to the top menu bar and click the **Statistics** button page of BIS and select **Exchange rates**. This directs the user to the portal page.

Step 2. Click the **Effective exchange rates**[3] and then click the **Data** button to access this data page.

Step 3. Go to the **Filters** and select from the pull-down menu. For example, select **Daily** from **Frequency**, **Nominal** from **Type**, **Broad** from **Basket** and **G20** from **Reference area**.

Step 4. All 19 areas are filtered. For USA, the user can click the **Export** button on the top ban and a pop-up window appears, as shown in Fig. 1.7.

[3] For simplicity, it is https://data.bis.org/topics/EER/data

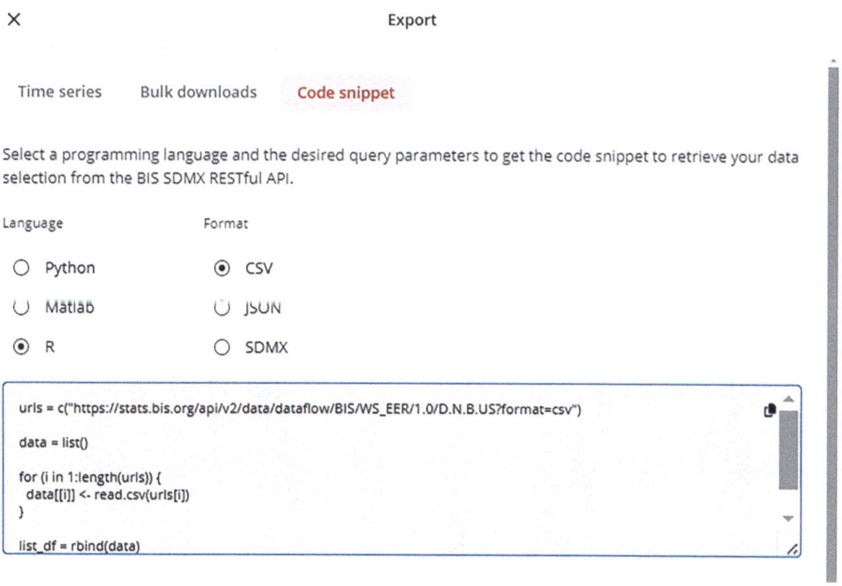

Fig. 1.8 The **Code snippet** button offers several choices for Language and Format. For Language, select R and for Format, select CSV

Step 5. Go to the **Code snippet** button, which has three choices of output formats and languages, as shown in Fig. 1.8. The URLs that are generated are then copied and pasted into the Code Chunk.

The following commands are then written in the Code Chunk as:

```
DNB_urls="https://stats.bis.org/api/v2/data/dataflow/BIS/WS_
EER/1.0/D.N.B.US?format=csv"
DNB=read.csv(neer_urls)[,c("TIME_PERIOD","OBS_VALUE")]
    DNB_us=xts::as.xts(DNB[-1],
                       order.by=as.Date(DNB[,1]))
    colnames(DNB_us)="US"
    DNB_us=na.omit(DNB_us)
    dev.new();plot(DNB_us)
```

The last line plots the retrieved data, and the result is shown in Fig. 1.9.

For the G20, in the Reference area, the selection is made and the URL generates the following content:

```
"https://stats.bis.org/api/v2/data/dataflow/BIS/WS_
EER/1.0/D.N.B.AR+AU+BR+CA+CN+DE+FR+GB+ID+IN+IT+JP+KR+MX+RU+SA+TR+
US+ZA?format=csv"
```

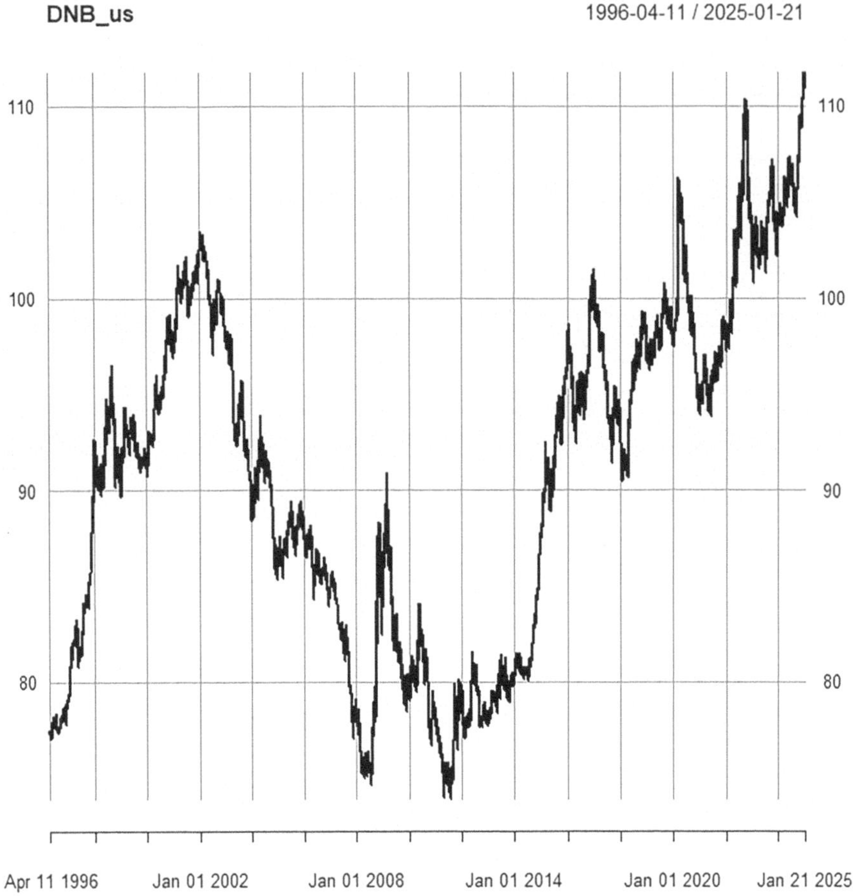

Fig. 1.9 The daily time series for Nominal Broad Effective Exchange Rates for the USA. (Source: Bank of International Settlement)

This URL cannot be directly used to download relevant series: they must be separated into three segments as:

```
main="https://
stats.bis.org/api/v2/data/dataflow/BIS/WS_EER/1.0/D.N.B."
end="?format=csv"
areas=c("AR","AU","BR","CA","CN","DE","FR","GB","ID","IN","IT",
"JP","KR",
        "MX","RU","SA","TR","US","ZA")
```

These are pasted together and named in terms of AREA below. A loop is then initiated as follows:

```
AREA=paste0(main, areas, end)
DNB_G20=NULL
for (i in AREA) {
tmp=read.csv(i)[,c("TIME_PERIOD","OBS_VALUE")]

tmp=xts::as.xts(tmp[-1],
                order.by=as.Date(tmp[,1]))
DNB_G20=cbind(DNB_G20, tmp)
print(i)
}
colnames(DNB_G20)=areas
```

This loop retrieves 19 daily nominal broad effective exchange rates series sequentially and returns a `xts` time series object `DNB_G20`. The raw data table has many unused columns so the `read.csv` component in the loop subsets two columns: `"TIME_PERIOD" and "OBS_VALUE"`, which respectively correspond to timestamp and data value.

By the same method, the bilateral exchange rates and data on BIS can be downloaded.

This procedure can be simplified using `JFE` wrapper `getEER`:

```
output=JFE::getEER(Areas=c("US","JP"),
                   Freq="Monthly",
                   Type="Real",
                   Basket="Broad")
```

where

`Areas`: The currency country for effective exchange rates (EER) for BIS, which allows many countries. The defaults are "US" and "JP."

`Freq`: The EER has two types: "Daily" and "Monthly." "Daily" works only for "Nominal" EER. The default is "Monthly."

`Type`: The EER has two types: "Nominal" and "Real." The default is "Real."

`Basket`: The basket for EER has two types: "Broad" and "Narrow." The default is "Broad."

The retrieved data frame is structured using timestamps, but it has no specific format. Users can transform it into either `timeSeries` or `xts`. Finally, `head(output)` views the data that is retrieved.

To retrieve the BIS data, the correct country ID is input to the `Areas` argument. Table 1.1 presents all of the choices.

Table 1.1 The BIS country ID

Country names	BIS.ID	Country names	BIS.ID
Algeria	DZ	Korea	KR
Argentina	AR	Latvia	LV
Australia	AU	Lithuania	LT
Austria	AT	Luxembourg	LU
Belgium	BE	Malaysia	MY
Bosnia & Herzegovina	BA	Malta	MT
Brazil	BR	Mexico	MX
Bulgaria	BG	Morocco	MA
Canada	CA	Netherlands	NL
Chile	CL	New Zealand	NZ
China	CN	North Macedonia	MK
Chinese Taipei	TW	Norway	NO
Colombia	CO	Peru	PE
Croatia	HR	Philippines	PH
Cyprus	CY	Poland	PL
Czechia	CZ	Portugal	PT
Denmark	DK	Romania	RO
Estonia	EE	Russia	RU
Euro area	XM	Saudi Arabia	SA
Finland	FI	Seia	RS
France	FR	Singapore	SG
Germany	DE	Slovakia	SK
Greece	GR	Slovenia	SI
Hong Kong SAR	HK	South Africa	ZA
Hungary	HU	Spain	ES
Iceland	IS	Sweden	SE
India	IN	Switzerland	CH
Indonesia	ID	Thailand	TH
Ireland	IE	Turkey	TR
Israel	IL	United Arab Emirates	AE
Italy	IT	United Kingdom	GB
Japan	JP	United States	US

1.2.4 `JFE::getFrench` Retrieves Asset Pricing Data

Dr. Kenneth French's data library[4] is an authoritative data source for asset pricing. It is popular with asset pricing economists and practitioners. The direct way is to point the cursor to the required data and right click to **Save As**. In French's data library, the asset pricing data is generally classified into two

[4] https://mba.tuck.dartmouth.edu/pages/faculty/ken.french/data_library.html

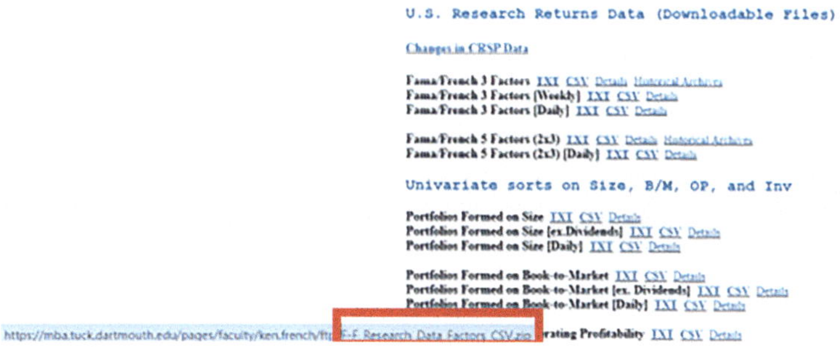

Fig. 1.10 Hover over the hyperlink to show the name that is stored in the database

types: factors and portfolios. The R package `JFE` offers two functions to retrieve these: `getFrench.Factors` and `getFrench.Portfolios`. Examples are explained below:

To retrieve data, the downloaded filename is copied to the function. To illustrate, Fig. 1.10 shows that to get the Fama/French 3 Factors CSV, hovering over the hyperlink causes the filename to appear on the bottom.

The name, `F-F_Research_Data_Factors`, is then pasted to the function to obtain the data as:

```
output=JFE::getFrench.Factors(filename="F-F_Research_Data_Factors")
```

Two objects are stored in the `output$table.names`,

```
> output$table.names

[1] "Monthly Factors"  "Annual Factors: January-December"
```

To retrieve the data, the process uses:

```
output$data$`Monthly Factors`
output$data$`Annual Factors: January-December'
```

Since `output$data` is a list, `output$data[[1]]` can also be used to retrieve monthly factors: the number in the double brackets corresponds the order for `output$table.names`.

Both frequencies are shown in Fig. 1.11. This output is not yet a time series, and readers are encouraged to transform it into a time series object, if help is needed, please check Chapter Code Chunk.

`output$file.name` returns the name of this data file.

By the same method, the portfolio data is retrieved using the following commands:

Dates	Mkt.RF	SMB	HML	RF
192607	2.96	-2.56	-2.43	0.22
192608	2.64	-1.17	3.82	0.25
192609	0.36	-1.40	0.13	0.23
192610	-3.24	-0.09	0.70	0.32
192611	2.53	-0.10	-0.51	0.31
192612	2.62	-0.03	-0.05	0.28
...	...	...	...	...
202407	1.24	6.80	5.74	0.45
202408	1.61	-3.55	-1.13	0.48
202409	1.74	-0.17	-2.59	0.40
202410	-0.97	-1.01	0.89	0.39
202411	6.51	4.63	-0.05	0.40
202412	-3.17	-2.73	-2.95	0.37

Dates	Mkt.RF	SMB	HML	RF
1927	29.47	-2.04	-4.54	3.12
1928	35.39	4.51	-6.17	3.56
1929	-19.54	-30.70	11.67	4.75
1930	-31.23	-5.17	-11.54	2.41
1931	-45.11	3.70	-13.95	1.07
1932	-9.39	4.40	11.11	0.96
...	...	...	...	...
2019	28.28	-6.13	-10.37	2.15
2020	23.66	12.72	-46.10	0.45
2021	23.57	-3.78	25.39	0.04
2022	-21.58	-7.04	25.97	1.43
2023	21.69	-3.28	-13.70	4.95
2024	19.78	-11.13	-9.10	5.26

Fig. 1.11 Fama-French's Factor Data, from 1926/7 to 2024/12; the publication lags calendar time by about 3 months. (**a**) Monthly factors for excess market returns, small-big size factor, high-low leverage factor, and risk-free rate, and (**b**) annual factors for excess market returns, small-big size factor, high-low leverage factor, and risk-free interest rate

```
output2=JFE::getFrench.Portfolios(filename="Portfolios_Formed_on_ME")
```

There are six tables in this file:

There are six tables in this file:

```
> output2$table.names
[1] "Value Weight Returns -- Monthly"
[2] "Equal Weight Returns -- Monthly"
[3] "Value Weight Returns -- Annual from January to December"
[4] "Equal Weight Returns -- Annual from January to December"
[5] "Number of Firms in Portfolios"
[6] "Average Firm Size"
```

Figure 1.12 shows the first "Value Weight Returns—Monthly".
For users' convenience, seven filenames for factors are listed as:

Dates	<= 0	Lo 30	Med 40	Hi 30	Lo 20	Qnt 2	Qnt 3	Qnt 4	Hi 20	Lo 10	Dec 2	Dec 3	Dec 4	Dec 5	Dec 6	Dec 7	Dec 8	Dec 9	Hi 10
192607	-99.99	0.14	1.54	3.42	0.37	0.48	1.68	1.41	3.67	-0.12	0.52	-0.05	0.82	1.39	1.89	1.62	1.29	3.53	3.71
192608	99.99	3.19	2.73	2.91	2.21	3.58	3.70	1.50	3.07	1.13	2.55	4.00	3.32	2.76	4.41	1.54	1.49	0.61	3.79
192609	-99.99	-1.73	-0.88	0.80	-1.39	-1.25	0.07	-0.23	0.81	0.59	-2.00	-2.01	-0.77	-0.01	0.13	-1.93	0.74	-0.77	1.25
192610	-99.99	-2.94	-3.26	-2.79	-2.56	-3.99	-2.65	-3.36	-2.74	-4.29	-2.01	-3.25	-4.45	-3.02	-2.39	-3.55	-3.26	-3.36	-2.56
192611	99.99	0.38	3.73	2.74	-0.95	3.03	3.50	3.25	2.71	-3.28	-0.23	0.08	4.90	3.56	3.45	3.60	3.05	3.86	2.40
192612	-99.99	4.15	1.66	3.04	2.45	3.47	1.37	2.81	3.00	-2.49	3.93	5.60	2.22	1.33	1.40	1.81	3.37	3.11	2.97
...	...	...	...	...	...	...	...	...	...	...	...	...	...	...	...	...	...	...	...
202407	-99.99	11.75	7.69	1.05	12.24	10.70	8.07	5.59	0.79	9.95	13.80	11.21	10.39	7.83	8.23	6.08	5.33	3.47	0.43
202408	-99.99	-3.17	-0.23	2.39	-2.92	-1.48	-1.05	1.50	2.42	-2.94	-2.92	-3.45	-0.30	-1.22	-0.93	0.84	1.84	1.77	2.51
202409	-99.99	-0.02	1.81	2.21	-0.45	0.34	1.89	2.27	2.21	0.48	-1.05	0.46	0.27	2.25	1.64	2.38	2.21	2.41	2.19
202410	-99.99	0.15	-0.73	-0.59	0.33	-1.12	-1.22	-0.57	-0.56	0.61	0.14	-0.05	-1.74	-0.57	-1.66	0.32	-1.03	-0.20	-0.62
202411	-99.99	10.70	10.43	6.53	9.90	12.06	10.89	8.73	6.40	10.78	9.30	11.58	12.34	12.85	9.55	9.08	8.55	9.31	6.00
202412	-99.99	-6.67	-7.58	-2.34	-5.36	-7.96	-8.21	-6.78	-2.06	-3.22	-6.82	-8.10	-7.87	-7.84	-8.47	-6.69	-6.83	-7.33	-1.31

Fig. 1.12 Data table for value weight returns—monthly frequency, where -99.99 denotes missing values

```
c("F-F_Research_Data_Factors_weekly",
  "F-F_Research_Data_Factors_daily",
  "F-F_Research_Data_5_Factors_2x3_daily",
  "F-F_Momentum_Factor_daily",
  "F-F_Research_Data_Factors",
  "F-F_Research_Data_5_Factors_2x3",
  "F-F_Momentum_Factor" )
```

To save space, 32 portfolio filenames are listed in the Chapter Code Chunk.

1.3 Compute Out-of-Sample Forecasts: Onestep and Multistep

To retrieve information on monthly inflation for June 2025, forecasting determines the future starting from the current day. Onestep forecast concerns July, and multistep concerns more months.

Onestep and multistep predictions are determined by the method that is used to specify predictive modeling. To obtain a onestep forecast, a predictive equation with exogenous regressors is used:

$$y_t = a + b\mathbf{X}_{\mathbf{t-1}} + e_t \tag{1.1}$$

This equation is applied to the 1947M1-2025M6 monthly inflation rates. When the parameters are estimated, the conditional mean is a onestep in-sample model fit, and the out-of-sample prediction for 2025M7 is computed as:

$$\hat{y}_{t+1} = a + b\mathbf{X}_{\mathbf{t}} \tag{1.2}$$

When the actual value for y_{t+1} is realized, the prediction error is calculated as $\hat{e}_{t+1} = y_{t+1} - \hat{y}_{t+1}$, and $\hat{e}_{t+1} \sim IID\left(0, \sigma^2\right)$. According to $E\left[\hat{e}_{t+1} | \mathbf{X}_{\mathbf{t}}\right] = 0$ and $\hat{y}_{t+1} = E\left[y_{t+1} | \mathbf{X}_{\mathbf{t}}\right]$, Eq. (1.2) is rewritten as a onestep forecast as:

$$\begin{aligned}\hat{y}_{t+1} &= E\left[a + b\mathbf{X}_t + \hat{e}_{t+1} | \mathbf{X}_t\right] \\ &= a + b\mathbf{X}_t + E\left[\hat{e}_{t+1} | \mathbf{X}_t\right] \\ &= a + b\mathbf{X}_t\end{aligned}$$

When more actual values of $\mathbf{X}_t$ are realized, Eq. (1.1) is updated and Eq. (1.2) is used to predict the next month.

Sometimes, low-frequency forecasting, especially an economic forecast, has a further horizon, such as 6 months, so a multistep forecast is necessary. However, Eq. (1.1) does not work for multistep, because it is limited to the actual values of $\mathbf{X}_t$. Therefore, a predictive equation with a lagged endogenous variable is used. This is also known as autoregression AR:

$$y_t = a + \phi y_{t-1} + e_t \tag{1.3}$$

Under AR(1) for Eq. (1.3), multistep forecasts are performed as:

$$\begin{aligned}&\hat{y}_{t+1} = \phi y_t, && h = 1 \\ &\hat{y}_{t+2} = \phi \hat{y}_{t+1}, && h = 2 \\ &\ldots \\ &\hat{y}_{t+6} = \phi \hat{y}_{t+5}, && h = 6\end{aligned}$$

These multistep forecasts are computed step-by-step or recursively obtained. The recursive strategy for multistep forecasts is also named "iterated multi-step" (IMS, see Chevillon, 2007) forecasting, because these are obtained dynamically by repeatedly iterating the model and by inputting the missing future values with their respective forecasts.

A disadvantage of the recursive multistep forecast is its sensitivity to estimation errors because estimated values are used instead of actual values as the length of the forecast increases. In spite of these limitations, the recursive strategy is used to forecast many real-world time series by using different machine learning models, such as RNN (Saad et al., 1998) and nearest-neighbors (McNames, 1998; Bontempi et al., 1999).

1.4 Forecast Accuracy Measures

If y and $\hat{y}$, respectively, denote the actual and predicted values, the prediction accuracy in terms of prediction error $e_t = y_t - \hat{y}_t$ is calculated using four exact measures. There are many loss functions. Makridakis (1993) studied the stability of predictors and showed that the Mean Absolute Prediction Error (MAPE) is relatively stable. Hyndman and Koehler (2006) used Mean Absolute Scaled Errors (MASE). If there is little data, there is no difference in the processing for scaling so the commonly used loss metrics are described as:

1. Two scale-dependent loss functions

   The first is RMSE (root mean squared errors): RMSE = $\sqrt{mean\left(e_t^2\right)}$

   where *mean* is the average and e_t is the prediction error.

   For MAE (mean absolute error): MAE = $mean(|e_t|)$,

2. Percentage errors

   If $p_t = \dfrac{\hat{y}_t - y_t}{y_t}$, the MAPE(mean absolute percentage error) is defined as: MAPE = $mean(|p_t|)$

   MAPE is subject to the raw data. If the raw data is about a sale quantity, MAPE tends to infinity if there is no transaction on a single day. Caution must be used to interpret the results.

3. Theil's U

   Theil's U is defined as: $\sqrt{\dfrac{\Sigma\left(\hat{y}_t - y_t\right)^2}{\Sigma y_t^2}}$, which has similar disadvantages to MAPE.

4. Autocorrelation of errors at lag 1

   ACF(1) (autocorrelation of errors at lag 1) calculates the persistence of prediction errors, which is defined by ϕ from the regression below:

$$e_t = \phi e_{t-1} + \varepsilon_t$$

   The ACF(1) coefficient ϕ also measures the speed at which the residuals decay over time. If the value is close to 1, the prediction errors are more persistent, so the model's predictive performance is not good and the prediction errors decay slowly to zero, whereas a value of 0.2 implies the prediction errors decay quickly to zero.

In addition, if ρ denotes the correlation coefficient for a forecasting horizon is *h*, RMSE can be decomposed into three components:

$$\Sigma\frac{\left(\hat{y}_t - y_t\right)^2}{h} = \left(\frac{\Sigma\hat{y}_t}{h} - \overline{y}\right)^2 + \left(s_{\hat{y}} - s_y\right)^2 + 2\left(1-\rho\right)s_{\hat{y}}s_y$$

$$1 = \frac{\left(\frac{\Sigma\hat{y}_t}{h} - \overline{y}\right)^2}{\Sigma\frac{\left(\hat{y}_t - y_t\right)^2}{h}} + \frac{\left(s_{\hat{y}} - s_y\right)^2}{\Sigma\frac{\left(\hat{y}_t - y_t\right)^2}{h}} + \frac{2\left(1-\rho\right)s_{\hat{y}}s_y}{\Sigma\frac{\left(\hat{y}_t - y_t\right)^2}{h}}$$

$$\frac{\left(\frac{\Sigma\hat{y}_t}{h} - \overline{y}\right)^2}{\Sigma\frac{\left(\hat{y}_t - y_t\right)^2}{h}} : \text{Bias Proportion}$$

$$\frac{\left(s_{\hat{y}} - s_y\right)^2}{\sum \frac{\left(\hat{y}_t - y_t\right)^2}{h}} : \text{Variance Proportion}$$

$$\frac{2\left(1-\rho\right)s_{\hat{y}}s_y}{\sum \frac{\left(\hat{y}_t - y_t\right)^2}{h}} : \text{Covariance Proportion}$$

The three components sum to 1, so:

(a) **Bias proportion** measures how far the predicted value is from the true value.
(b) **Variance proportion** measures how much of the predicted variation is derived from real data.
(c) **Covariance proportion** measures the remaining unsystematic prediction error.

To assess a good forecast in terms of RMSE, the first two components must be small so that most of the prediction errors are concentrated in the third covariance proportion. To compare two prediction errors to infer forecasting performance, the Diebold–Mariano test (Diebold & Mariano, 1995) is used. This uses an equality test with two correlated samples. The following chapters show when it is used.

Chapter 2
Predictive Time Series Modeling

To implement a statistical prediction is to identify its central tendency or pattern; namely, location and scale. Examples are the parametric (conditional) mean-variance or the nonparametric median-IQR. Forecasting time series also involves identifying patterns. If the patterns are regular, the movement can be modeled using deterministic dummies. Figure 2.1 shows the data for real GDP per capita for the United States, which is decomposed into seasonal, trend, and irregular components using moving averages. There are three components of a low-frequency economic time series:

1. Trend: A long-term trend.
2. Cycle: The cycle effect is the periodic change for the time series, such as seasonality or calendar effects.
3. Irregular random variations: The random distractor that remains after the other two components are removed.

These stylized facts are common in low-frequency economic time series, but there are generally no obvious cycles for high-frequency data, such as daily, weekly, or intraday data.

Figure 2.1 shows that modeling time series using trending and seasonal dummies captures most of the movement but real economic time series contain unknown stochastic components, which cannot be identified using dummies. To contrast with a machine learning algorithm, this chapter initially considers statistical baseline models.

Equation (1.1) implements a onestep forecast but it cannot perform more step forecasts ahead (multistep) unless a forecasting model is constructed for $\mathbf{X}_t$ to generate forecasts ahead. The AR for Eq. (1.3) performs onestep and multistep predictions. In terms of economic theory, (1.1) is important for policy diagnosis and

T.-w. Ho, *Time Series Forecasting using Machine Learning*,
https://doi.org/10.1007/978-3-031-97946-0_2

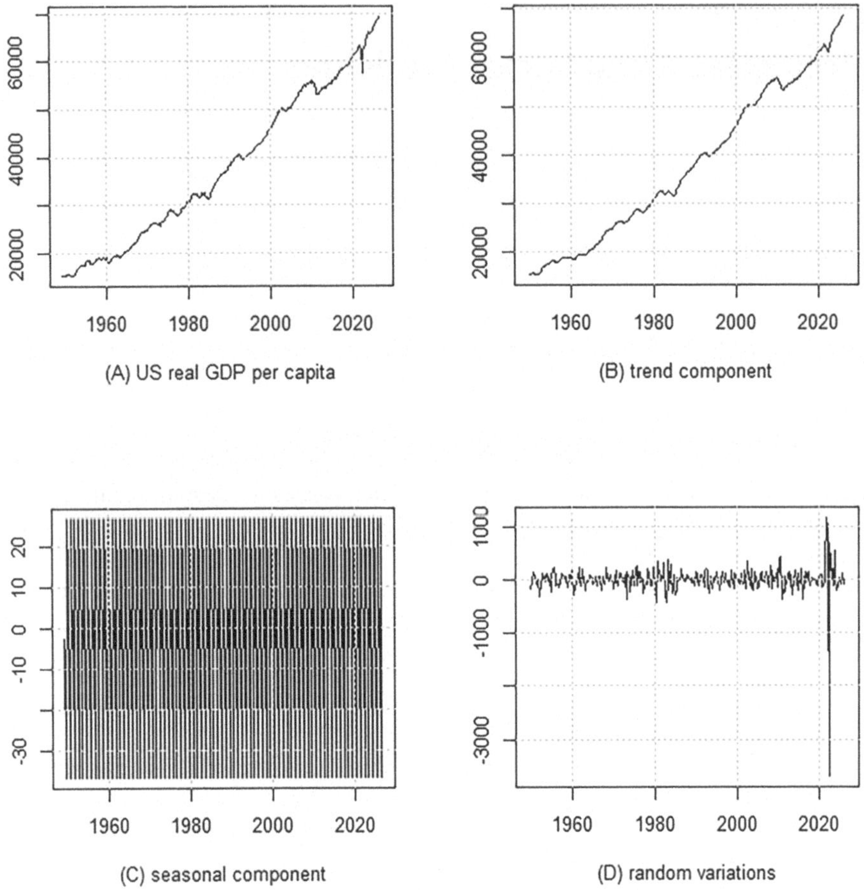

Fig. 2.1 Features of US real GDP per capita, 1947Q1–2024Q3: (**a**) US real GDP per capita, (**b**) trend component, (**c**) random variation, and (**d**) seasonal component

justification, so from the practitioners' viewpoint, onestep prediction with lower frequency, such as quarterly data, may be better.

The disadvantage is that at time t, the data $\mathbf{X_t}$ is usually unavailable, so it is almost impossible to forecast y_{t+1}. In reality, in February, there is no February data to forecast March and complete a onestep prediction. If the date on which this paragraph is written is 2025/2/15, the CPI is 2024/12: not January.

The practice examples in this and the other chapters are based on multistep or dynamic forecasting. Other applications are discussed in Chap. 5.

2.1 ARIMA

2.1.1 Model Explanations

The ARIMA model uses autocorrelation. The current expected price is linearly related to the previous price. The limitation of the ARIMA model is that choosing appropriate values for the AR and MA terms requires consideration of the ACF and PACF graphs and the theoretical concept of statistical modeling in ARIMA implementation must be understood. If the conditions for choosing appropriate values for the AR and MA terms are met, ARIMA is a good model, as detailed in Chapter 2 of the work by Tsay (2010).

For a univariate time series y_t, without exogeneous explanatory variables, y_t must be predicted. Statistical prediction relies on expected values, so the simple sample mean ($\bar{y}$) can be used:

$$\bar{y} = \frac{\sum_{t=1}^{T} y_t}{T}$$

The simple sample average is not a good expectation but a time series can predict forwards using its own past values. This is an autoregressive process (AR):

$$y_t = \alpha + \phi_1 y_{t-1} + \phi_2 y_{t-2} + \cdots + \varepsilon_t \quad (2.1)$$

The error in the time series is generally called the innovation process, which describes a disturbance adjustment term. If y_t is a stationary time series, it satisfies the following conditions:

$$E[\varepsilon_t] = 0, \forall t \quad (2.2a)$$

$$E[\varepsilon_t^2] = \sigma^2, \forall t \quad (2.2b)$$

$$E[\varepsilon_s \varepsilon_t] = 0, \forall s \neq t \quad (2.2c)$$

These three conditions are the same as the requirements for the standard regression model. Equation (2.1) is generally written as a p-order lag, which is abbreviated as AR(p), as:

$$y_t = \alpha + \phi_1 y_{t-1} + \phi_2 y_{t-2} + \cdots + \phi_p y_{t-p} + \varepsilon_t \quad (2.3)$$

Ideally, the error term ε_t for Eq. (2.3) is white noise, which satisfies the three conditions for Eqs. (2.2a)–(2.2c). If $\boldsymbol{L}$ is the lag operator, and $\mathbf{L}y_t = y_{t-1}$ and $\mathbf{L}^k y_t = y_{t-k}$, Eq. (2.3) can be expressed as:

$$\phi(L) y_t = \alpha + \varepsilon_t \quad (2.4a)$$

$$\phi(L)=1-\phi_1L-\phi_2L^2-\cdots-\phi_pL^p \tag{2.4b}$$

The statistical properties of the AR(p) time series in Eq. (2.3) are determined by the calculated coefficients φ_1, φ_2,... φ_p. For example, the roots of the polynomial in (2.4b) determine whether it is stationary. The roots of the p-order polynomial of Eq. (2.4b), represented as the roots of the polynomial $\varphi(z)$, which determine whether this polynomial is a convergent series, are decomposed as:

$$\phi(z)=(1-\alpha_1z)(1-\alpha_2z)\cdots(1-\alpha_pz) \tag{2.5}$$

Therefore, a stationary AR(p) convergent time series, such as Eq. (2.5) satisfies $|\alpha_k| < 1$ for all $k\in\{1,2\ldots p\}$. AR(1) is used to explain this property. For simplicity, the following AR(1) without intercept is used:

$$y_t=\phi y_{t-1}+\varepsilon_t,\ y_t\sim(\mu,\gamma_0) \tag{2.6}$$

The step-by-step substitution of Eq. (2.6) is:

$$\begin{aligned}
&t=1\Rightarrow y_1=\phi y_0+\varepsilon_1\\
&t=2\Rightarrow y_2=\phi y_1+\varepsilon_2\Rightarrow y_2=\phi(\phi y_0+\varepsilon_1)+\varepsilon_2=\phi^2y_0+\phi\varepsilon_1+\varepsilon_2\\
&\vdots\\
&y_t=\phi^ty_1+\sum_{j=0}^{t-1}\phi^j\varepsilon_{t-j},t=1,\ldots,T
\end{aligned}$$

where y_0 is the initial value, so the expanded general formula $y_t=\phi^ty_1+\sum_{j=0}^{t-1}\phi^j\varepsilon_{t-j},t=1,\ldots,T$, implies that if $|\phi| < 1$, the effect of the initial period converges and the further the error value, the smaller is the effect on the current value for y_t. If $|\phi| \geq 1$, these influences diverge, so $|\phi| < 1$ is a stationary condition. The algebraic proof is:

If $y_t\sim(\mu,\gamma_0)$, by $y_t=\phi^{t-1}y_1+\sum_{j=0}^{t-2}\phi^j\varepsilon_{t-j}$, then:

$$Ey_t=\phi^tEy_1+\sum_{j=0}^{t-2}\phi^jE\varepsilon_{t-j}=\phi^tEy_1+0$$

$$\Rightarrow\mu=\phi^t\mu \tag{2.7}$$

Secondly, according to the definition of variance:

$$\begin{aligned}
\gamma_0&=E(y_t-\mu)^2=Ey_t^2-\mu^2\\
&=\gamma_0+\mu^2=Ey_t^2\\
&=E(\phi y_{t-1}+\varepsilon_t)^2=\phi^2E(y_{t-1}^2)+\sigma^2
\end{aligned}$$

$$\Rightarrow \gamma_0 + \mu^2 = \phi^2 \left(\gamma_0 + \mu^2\right) + \sigma^2 \tag{2.8}$$

Equation (2.7) holds true only if $\mu = 0$ or $\phi = 1$. However, $\phi = 1$ does not allow γ_0 to have a finite solution, so $\mu = 0$ or $\phi \neq 1$.

By Eq. (2.8), $\gamma_0 = \gamma_0\phi^2 + \sigma^2 \Rightarrow 1 - \phi^2 = \frac{\sigma^2}{\gamma_0}$

$$\Rightarrow \phi^2 = \frac{\gamma_0 - \sigma^2}{\gamma_0}, \text{which proves the stationary condition } \phi < 1$$

The moving average (MA) process for y_t, which is usually expressed as MA(q), can be written as:

$$y_t = \alpha + \varepsilon_t + \theta_1\varepsilon_{t-1} + \theta_2\varepsilon_{t-2} + \cdots + \theta_q\varepsilon_{t-q} \tag{2.9}$$

This MA(q) process still has a white noise error term: $\varepsilon_t \sim IID(0, \sigma^2)$, which also satisfies:

$$\mu = E\left[y_t\right] = \alpha$$

$$\text{variance}: \gamma_0 = \sigma^2 \left(1 + \sum_{j=1}^{q} \theta_j^2\right)$$

$$\text{co-variance}: \gamma_k = \sigma^2 \left(\theta_k + \sum_{j=k+1}^{q} \theta_j \theta_{j-k}\right) \text{for k} \leq \text{q and } \gamma_k = 0 \text{ for k} > \text{q}$$

An MA process represents y_t as the current and past error terms (white noise), and an AR process represents y_t as its past self. Therefore, this is an inverse operation so AR(p) and MA(q) are invertible and MA(q) can be written as an AR(p), where $p \to \infty$, as:

$$y_t - \varepsilon_t = \alpha + \theta_1\varepsilon_{t-1} + \theta_2\varepsilon_{t-2} + \cdots + \theta_q\varepsilon_{t-q}$$

Therefore, $\Rightarrow E[y_t | Y_{t-1}] = \alpha + \varepsilon_t + \theta_1\varepsilon_{t-1} + \theta_2\varepsilon_{t-2} + \cdots + \theta_q\varepsilon_{t-q}$

There are many interesting theoretical questions between AR(p) and MA(q), we will stop here for now and introduce a common property: AR(p). This can be written as MA(1), or MA(1) can be expanded into AR(p).

This property is often present in time series literature and it extends to the vector relationship between VAR(p) and VMA(1). The mathematical explanation is described as:

The time series y_t is written as a MA(1) process without intercept as:

$$y_t = \varepsilon_t + \theta_1\varepsilon_{t-1} \tag{2.10}$$

The invertibility condition allows the residuals to be expressed in terms of y_t and its lag:

$$\varepsilon_t = y_t - \theta_1 \varepsilon_{t-1} \tag{2.11}$$

Using the step-by-step substitution as in Eq. (2.6):

$$\varepsilon_2 = y_2 - \theta \varepsilon_1$$
$$\varepsilon_3 = y_3 - \theta \varepsilon_2 = y_3 - \theta \left(y_2 - \theta \varepsilon_1 \right) \Rightarrow \varepsilon_3 = y_3 - \theta y_2 + \theta^2 \varepsilon_1$$
$$\varepsilon_4 = y_4 - \theta \varepsilon_3 = y_4 - \theta \left(y_3 - \theta y_2 + \theta^2 \varepsilon_1 \right) \Rightarrow \varepsilon_4 = y_4 + (-\theta) y_3 + (-\theta)^2 y_2 + (-\theta)^3 \varepsilon_1$$
$$\ldots$$
$$\varepsilon_t = y_t - \theta y_{t-1} + \theta^2 y_{t-2} - \cdots + (-\theta)^{t-2} y_2 + (-\theta)^{t-1} \varepsilon_1 \tag{2.12}$$

Invertibility requires that the last error term on the right side of Eq. (2.12) disappears over time, so there is no residual term in the right middle. This implies that $-1 < \theta < 1$ or $|\theta| < 1$. Therefore, from Eq. (2.10) to Eq. (2.12), an MA(1) can be expanded into an AR(p) and $p \to \infty$ still holds.

In an AR(p) or MA(q) framework, the order of p or q is often very large, so a large number of parameter values must be calculated and there is a large number of lagged variables. Therefore, statisticians use the following method to create fewer orders:

$$y_t = \frac{\theta(L)}{\phi(L)} \varepsilon_t$$

A pattern is obtained:

$$\phi(L) y_t = \alpha + \theta(L) \varepsilon_t \tag{2.13}$$

Equation (2.13) has the ARMA(p, q) structure. For non-time series data, estimating the conditional expectation requires exogenous explanatory variables, i.e., X, to make the linear regression $E[y|X] = a + bX$ possible. Therefore, for cross-section data, estimating the conditional expectation can be difficult, so the sample expectation (mean) is crucial. Therefore, big data and data science are of great significance for this field of study.

For a time series, if there are no explanatory variables, the conditional expectation is estimated using its own past values to generate predictions. This model and packaged software are now generally used, but in the 70s of Box–Jenkins, it was a significant contribution.

Theoretical Exercise If Eq. (2.6) is changed to $\phi = 1$, so $y_t = y_{t-1} + \varepsilon_t$, $\varepsilon_t \sim IID(0, \sigma^2)$, derive the variance of y_t to explain its non-stationary property.

R Lab is now implemented. This simulates an AR(1) process with a DGP (data generating process) $y_t = 0.5y_{t-1} + \varepsilon_t$.

2.1.2 RLab. Simulation

Code Chunk. DGP:$y_t = 0.5y_{t-1} + \varepsilon_t$

```
1. ar1.sim0=arima.sim(n = 200, list(ar = 0.5))
2. ar1.sim1=ts(ar1.sim0, end=c(2024,12), freq=12)
3. ar1.sim2=timeSeries::as.timeSeries(ar1.sim1)
Plot
4. par(mfrow=c(2,1))
5. plot(ar1.sim1,
        col="steelblue",
        ylab="AR(1)",
        xlab="(A) Draw a ts() object")
        grid()
6. plot(ar1.sim2,
        col="steelblue",
        ylab="AR(1)",
        xlab="(B) Draw a timeSeries () object")
        grid()
7. rug(as.vector(ar1.sim2), ticksize = 0.01, side = 2, quiet
   = TRUE)
8. par(mfrow=c(1,1))
```

Line 1 simulates $y_t = 0.5y_{t-1} + \varepsilon_t$ with 200 observations and the next two lines transform this into respective time series objects. `arima.sim()` is the base R function that is used to simulate the ARIMA process. For the first declaration, $n = 200$ is the number of observations and the second declaration is the AR(1) coefficient.

Line 2: `ar1.sim1=ts(ar1.sim0, end=c(2021,8), freq=12)` Sampled data is used to create a time series ts() object. Monthly data is generated, The last month is December 2024 and this is named, `ar1.sim1`.

Line 3 converts the object of Line 2 into a time series object.

The remaining commands plot both formats, as shown in Fig. 2.2.

Simulation is used to determine time series properties.

For DGP ARMA(2,1): $y_t = 0.35y_{t-1} + 0.25y_{t-2} + \varepsilon_t - 0.15\varepsilon_{t-1}$,

```
arima.sim(n=200,
          list(ar=c(0.35,0.25),
               ma=c(-0.15)))
```

ARIMA(p, d, q) matches the `order=c(p, d, q)`. The middle term `d` is integrated to order d, which is the order of difference, to render it stationary. It defines a first-order difference, and both AR and MA orders are 0, which represents a pure I(1) non-stationary time series.

For an ARMA that is integrated of order 1, one more option is declared in the `list()` argument: `order=c(1,1,1)`, which pre-differences the series before generating the ARMA.

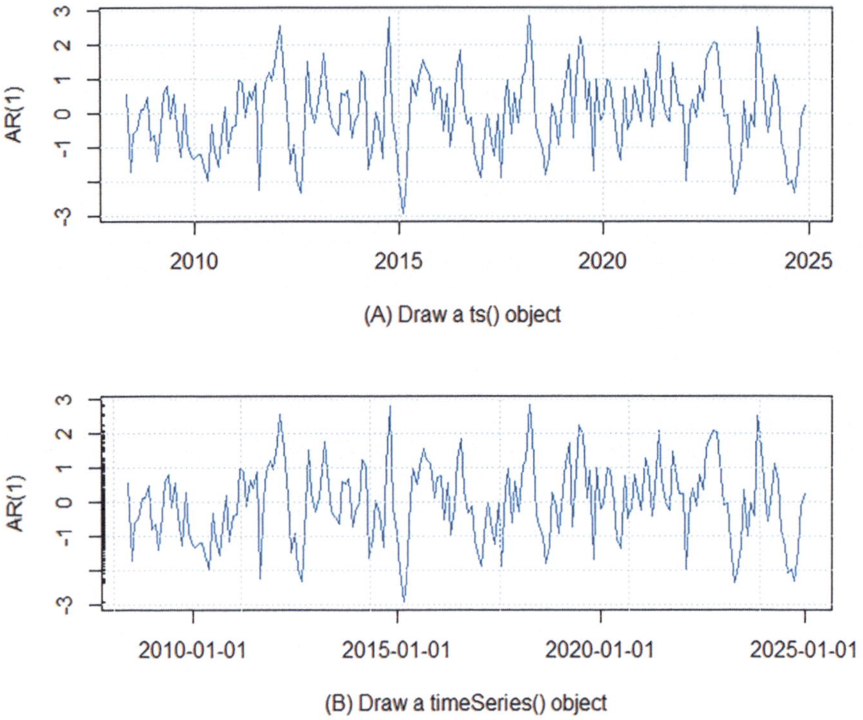

Fig. 2.2 Two time series objects with a DGP: $y_t = 0.5y_{t-1} + \varepsilon_t$

```
arima.sim(n = 600,
          list(order=c(1,1,1),
               ar=c(-0.135),
               ma=c(0.531)))
```

Finally, for an ARMA that is integrated of seasonal order 1, one more option is declared in the `list()` argument: seasonal=`c(1,1,1)`, which pre-differences the series before generating a seasonal ARMA.

```
arima.sim(n = 200,
          list(seasonal=c(1,1,1),
               sar=c(-0.135),
               sma=c(0.531)))
```

These two types of non-stationary time series are plotted in Figs. 2.3 and 2.4.

These two figures show that a simple differenced non-stationary time series visually represents a stochastic tend, so it is also known as a trend stationary. However, seasonal differenced stationarity visually lacks a stochastic trend. Analysis of low-frequency economic time series requires caution.

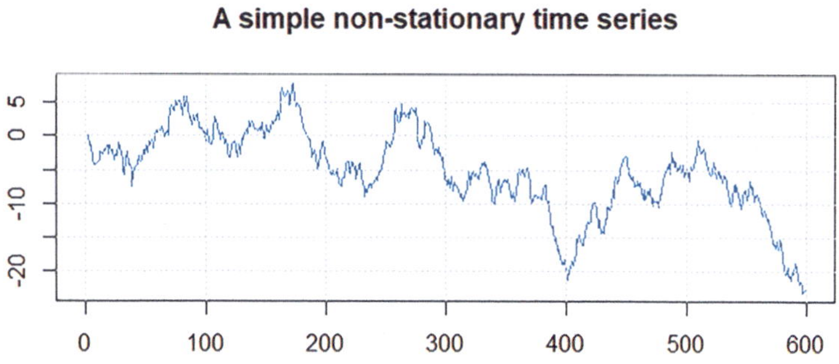

(A) ARIMA(0,1,0)

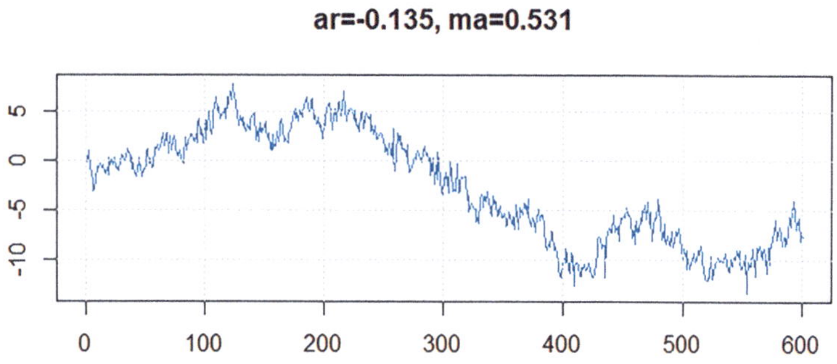

(B) ARIMA(1,1,1)

Fig. 2.3 Two I(1) non-stationary time series, where the time series is integrated to order 1 and generated using `arima.sim()`: (**a**) A I(1) non-stationary time series without ARMA and (**b**) An I(1) non-stationary time series with ARMA

This is flexible for a smaller standard deviation, such as $\sqrt{0.16}$, by specifying:

```
arima.sim(n=200,
          list(ar=c(0.5), ma=c(-0.25)),
          sd=sqrt(0.16))
```

To generate a financial time series with mild tail properties, randomly draw observations are drawn from the student-t distribution, which is the function **rt()**, as:

```
arima.sim(n=600,
            list(ar=c(0.88, -0.48),
                 ma=c(-0.22, 0.28)),
            rand.gen=function(n)sqrt(0.16)*rt(n,df=5))
```

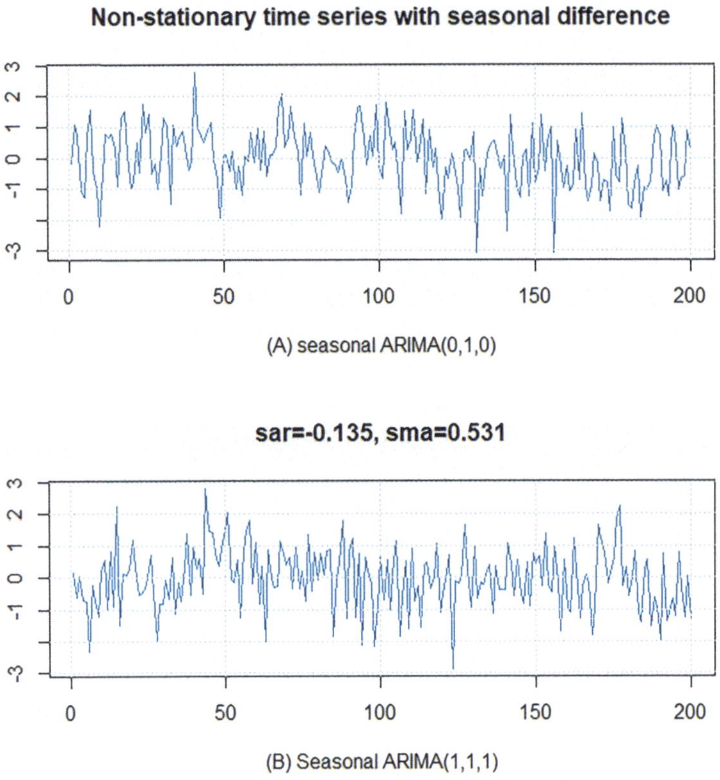

Fig. 2.4 Two seasonally non-stationary time series: the time series is generated using `arima.sim()` with seasonality: (**a**) is a non-stationary time series with seasonal difference and (**b**) is a non-stationary time series with seasonal ARMA and difference

There is a built-in standard normal distribution rand.gen=rnorm, so this is omitted. It is standard normal. At this time, a random variable with a degree of freedom of 5 and a variance of 0.16 is proposed:

```
sqrt(0.16) * rt(n, df = 5)
```

Figure 2.5 shows a time series plot. The nature of the data is not obvious if the time series movement is considered. The density plot or histogram must be plotted. Figure 2.6 uses $n = 3000$ samples and compares the density distribution for the standard normal distribution and the student-t distribution to show the long tail. Figure 2.7 uses a histogram. Using these two graphs, the tail properties can be categorized.

R Exercises

1. Adjust the parameters of the student t-distribution (such as degrees of freedom and variance). Which parameters have the greatest impact on the tail shape? Present this with a probability density plot (density plot) or histogram.

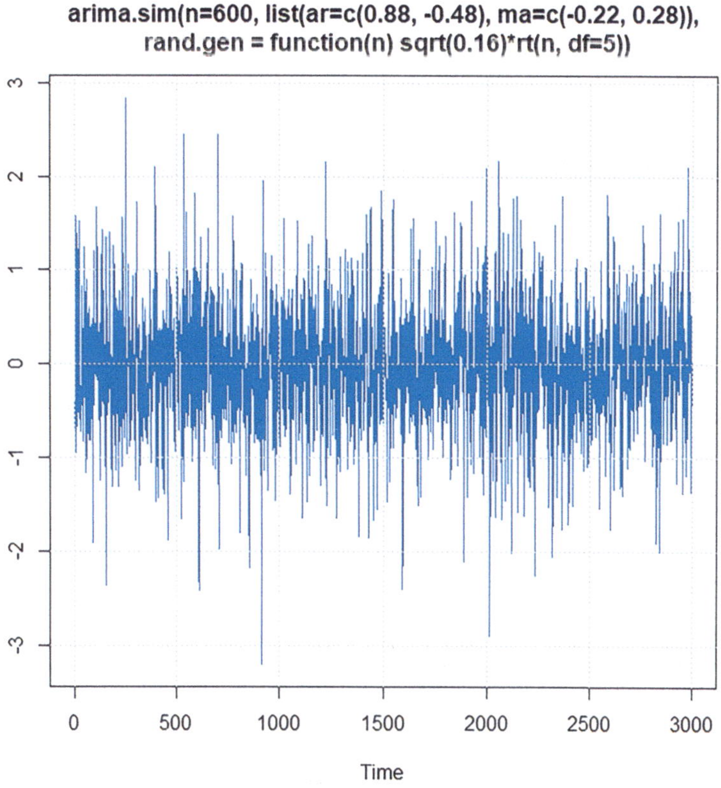

Fig. 2.5 Random sampling from the student-t distribution, `rt()`

2. Check the R documentation. Type the command `?Distributions` to view the Help documentation. Use other sampling functions (such as Weibull) and discuss the features that are visualized.

2.1.3 RLab. Forecasting US Core Inflation

This section concerns real-time series data. `getFed()` is used to download the US CPI, and the seasonally adjusted inflation is denoted as y.

In ARIMA(p, d, q), if the maximum values for (p, d, q) = (5, 2, 5), there are $6 \times 3 \times 6 = 108$ possible combinations. Finding the best order manually is very inefficient. The R package `forecast` features the `auto.arima()`, which performs all combinations and selects the best fit according to a specific indicator (default is AIC),as:

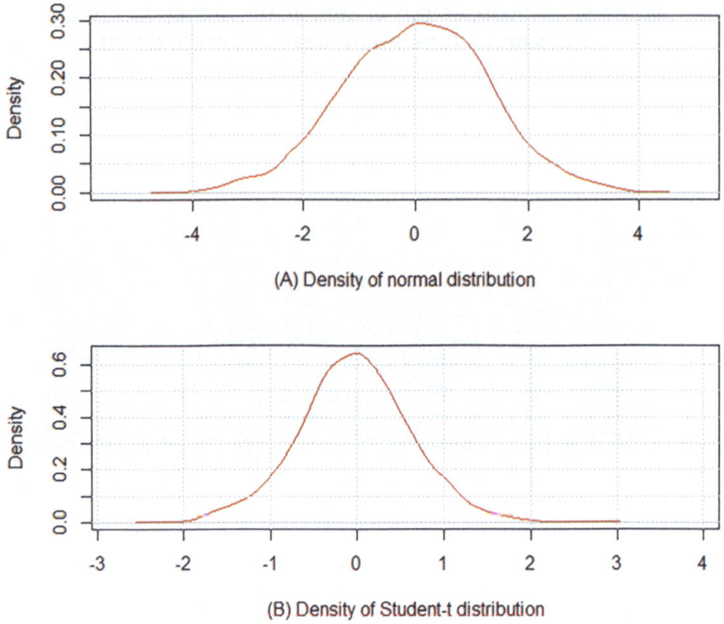

Fig. 2.6 Kernel density plot of two sampling distributions for which N = 3000 and bandwidth = 0.237. (**a**) Density of normal distribution and (**b**) density of student-t distribution

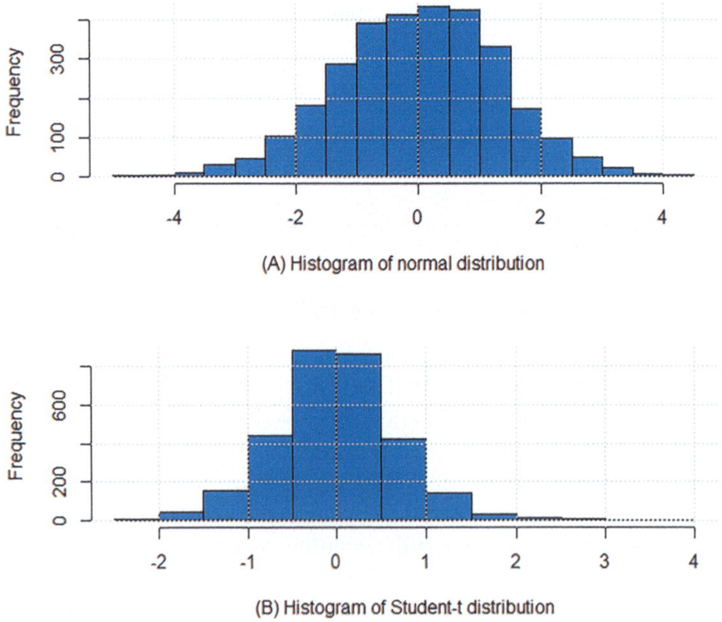

Fig. 2.7 Histogram of two sampling distributions for which N = 3000. (**a**) Normal distribution and (**b**) Student t-distribution for which bandwidth = 0.237

Fitting US Core Inflation

Code Chunk

```
library(forecast)
library(timeSeries)
auto.arima(y)
E_y=fitted(auto.arima(y))
plot(y, col="steelblue", xlab="Inflation rate of US, %", main="auto.arima():
ARIMA(2,1,2)(2,0,2)[12]", ylab="")
grid()
lines(E_y,col="red",lty=2)
rug(as.vector(y), ticksize = 0.01, side = 2, quiet = TRUE)
```

The result for auto.arima(y) is:

```
> out

Series: y

ARIMA(2,1,2)(2,0,2)[12]

Coefficients:

  ar1     ar2     ma1      ma2     sar1     sar2     sma1    sma2

0.192  0.7877  0.0300  -0.6118  0.5496  -0.0888  -1.6076  0.6395

s.e.  0.721  0.713  0.716  0.546  0.148  0.045  0.146  0.142


sigma^2 = 0.02662:  log likelihood = 305.38

AIC=-592.76   AICc=-592.53   BIC=-550.57
```

This result is ARIMA(2,1,1)(2,0,2)[12], which contains three sets of numbers:

(2,1,1) is a general structure, first-order difference, ARMA(2,1)
(2,0,2) is a seasonal structure with no seasonal difference and seasonal ARMA(2,2)
[12] is the data frequency, which is the monthly data

The data is fitted using the equation:

$$\Delta y_t = \phi_1 \Delta y_{t-1} + \phi_2 \Delta y_{t-2} + \varepsilon_t + \theta_1 \varepsilon_{t-1} + \phi_3 \Delta y_{t-12} + \phi_4 \Delta y_{t-24} + \theta_2 \varepsilon_{t-12} + \theta_3 \varepsilon_{t-24}$$

Fitting by `auto.arima()`, Fig. 2.8 plots two series together. The dashed line represents the fitted conditional expectations, and the solid line represents the actual data.

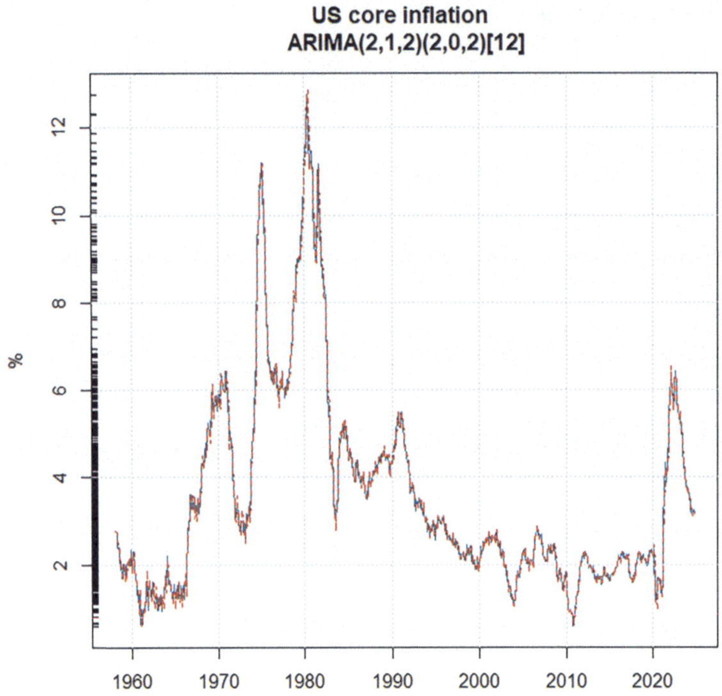

Fig. 2.8 US core inflation rate: dotted line shows the in-sample conditional expectation, as fitted by ARIMA(2,1,1)(2,0,2), so this is an I(1) process with ARMA(2,1) and seasonal ARMA(2,2)

Figure 2.8 plots actual data and an in-sample model fit (conditional mean) together. This looks good because the parameter estimates are optimized using all data points. To show predictability, the time is separated into two segments: the in-sample and the out-of-sample.

The in-sample estimation period ranges from 1957/1 to 2021/12, which is used to estimate the optimal values for the parameters. These optimal parameter values are used to predict the last 3 years as the out-of-sample months, ranging from 2022/1 to 2024/12. The in-sample and out-of-sample are also known as training and testing samples. These are separated by a vertical line in Fig. 2.9. The line is selected so that the out-of-sample features a reversal trend.

For ARIMA, `auto.arima()` is used directly to determine the optimal model, as shown in the following RLab:

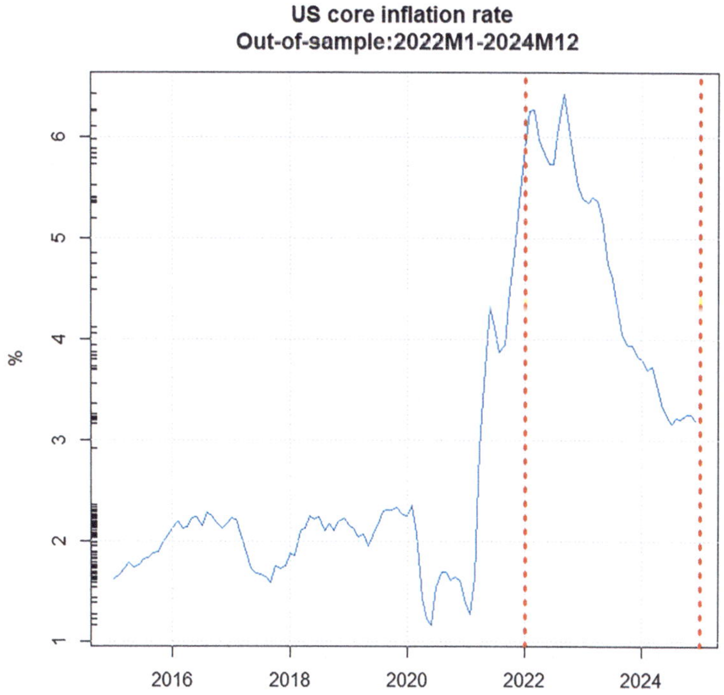

Fig. 2.9 A 10-year zoom-in US core inflation, separated by in-sample and out-of-sample periods. The in-sample estimation period ranges from 1957/1 to 2021/12, and this is used to estimate the optimal parameter values, which are used to predict the last 3 years as the out-of-sample months, ranging from 2022/1 to 2024/12

In-sample Estimation and Out-of-Ample Forecasts

Code Chunk

```
in_sample=window(y, end=c(2021,12))
out_sample=window(y, start=c(2022,1))
OUT1=auto.arima(in_sample)
Fcst1 = forecast(OUT1, h=length(out_sample))$mean
accuracy(Fcst1, x=out_sample)
```

The code chunk is divided into three steps:

Step 1. Separate the observations into two subsets:

```
in_sample=window(y, end=c(2021,12))
out_sample=window(y, start=c(2022,1))
```

Step 2. Estimation. `auto.arima()` is used to estimate the in_sample time series.

```
OUT1=auto.arima(in_sample, ic="aic")
```

To show the goodness-of-fit for the estimation, an in-sample fit is generated and `forecast::accuracy` is applied to it:

```
accuracy(fitted(OUT1),x=in_sample)
```

```
> accuracy(fitted(OUT1),x=in_sample)[,-c(1,4)]
    RMSE      MAE      MAPE      ACF1 Theil's U
  0.1646   0.1155   4.2655    0.0173    0.7481
```

Step 3. Multistep forecasting.

```
Fcst1 = forecast(OUT1, h=length(out_sample))$mean
```

The forecasting accuracy is calculated using `accuracy(Fcst1, x=out_sample)`, which is shown in Fig. 2.10. The results show that a good in-sample estimation does not guarantee good future forecasts. As Fig. 2.10 shows, the multistep forecasts feature good predictability in the initial months and the turning points are captured. After 10 months, from 2022M1, the forecasting gap increases.

Different information criteria can yield a different fitted order, but the parameters are optimized so the forecasting performance is not significantly different. For low-frequency economic data, seasonality and deterministic trends are important.

ARIMAx

ARIMAx is a version of ARIMA with exogeneous variables, which are usually known as independent variables, regressors, or covariates, including dummies. For seasonal dummies, the option `xreg` inside `auto.arima` is used. This can be extended to general independent variables (regressors) or covariates.

Code Chunk

```
Rhs1=seasonaldummy(in_sample)
OUT2=auto.arima(in_sample, ic="aic", xreg =Rhs1)
Rhs2= seasonaldummy(out_sample)
Fcst2 = forecast(OUT2, h=length(out_sample),
                        xreg =XREG2)$mean
```

To add a deterministic trend to `Rhs1`, a sequence is subjected to `cbind`, as:

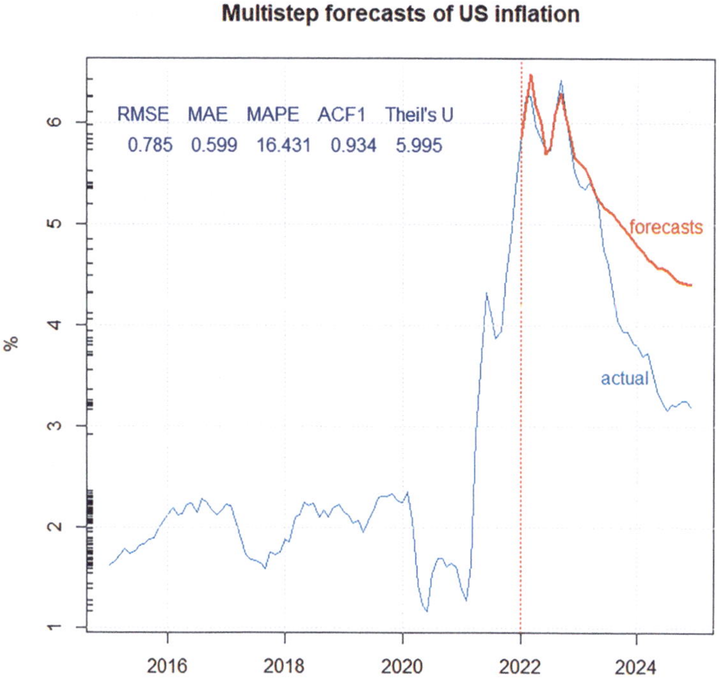

Fig. 2.10 Out-of-sample multistep forecasts for 2022M1–2024M12. The in-sample estimation period ranges from 1957/1 to 2021/12, which is used to estimate the optimal parameter values, and then these optimal values are used to predict the last 3 years as the out-of-sample months, ranging from 2022/1 to 2024/12. The results show that after 10 months from 2022M1, the forecasting gap increases. The Code Chunk file in this chapter contains the commands for this plot

```
Rhs1=cbind(seasonaldummy(in_sample),
                seq(in_sample))
```

Figure 2.11 plots the result, which shows that seasonal dummies are not a necessary addition. They decrease the accuracy of multistep forecasts, possibly because the seasonals are accounted for by seasonal ARMA, so unnecessary dummy variables decrease the accuracy of the prediction. Using a trend dummy will decrease the accuracy of a forecast, interested readers are encouraged to work on this.

Using the model structure of the previous RLab and adding a Fourier function to capture frequency-independent cycles, `xreg=fourier(in_sample, K=4)` is added to `auto.arima` to generate Fourier cycles. See Help for details about `fourier()`.

R Exercises

1. Retrieve the daily Dow Jones Index and select log of trading volume. Adjust the in-sample periods to 3 years and the out-of-sample to 20 days. Apply the SETAR

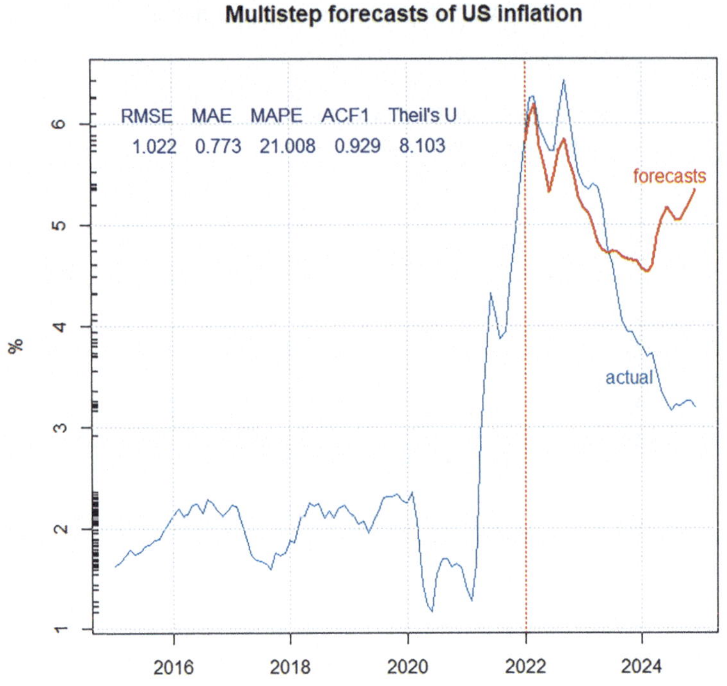

Fig. 2.11 Out-of-sample multistep forecasts: ARIMA with seasonal dummies. The result shows that after 10 months from 2022M1, the forecasting gap increases. Compared with Fig. 2.10, ARIMA with seasonal dummies is not a good model

and LSTAR models to generate an out-of-sample multistep forecast of the log volume.

2. Calculate the daily range volatility for the Dow Jones Index using the formula $\frac{(\ln \max - \ln \min)^2}{4 \ln 2}$ and then predict this range volatility as Exercise 1.

2.2 Nonlinear Transition Models: SETAR and LSTAR

2.2.1 Model Explanations

Two transition models are used: SETAR and LSTAR. The former is an acronym for *Self-Exciting Threshold AutoRegression*. This text focuses on implementation and prediction, and the theoretical details of these two methods are described in the work by Tsay (2010, Section 1 of Chapter). The SETAR(p) model uses Eq. (2.14):

$$y_t = \begin{cases} \mu_1 + \rho_{1,1} y_{t-1} + \cdots + \rho_{1,p1} y_{t-p1} + \varepsilon_t & \text{if} \quad x_{t-d} \le \theta_{m-1} \\ \mu_2 + \rho_{2,1} y_{t-1} + \cdots + \rho_{2,p2} y_{t-p2} + \varepsilon_t & \text{if} \quad \theta_{m-1} < x_{t-d} \le \theta_{m-2} \\ \vdots & \\ \mu_m + \rho_{m,1} y_{t-1} + \cdots + \rho_{m,pm} y_{t-pm} + \varepsilon_t & \text{if} \quad \theta_1 < x_{t-d} \end{cases} \tag{2.14}$$

where the parameters are:

m: Number of states
$\mu_1 \ldots \mu_m$: Intercept for each state
$p_{j,1} \ldots p_{j,m-1}$: Number of lags in state j
$\theta_1 \ldots \theta_{m-1}$: Threshold values
d: Number of delay terms for the transition variable
x_{t-d}: Threshold (transition) variable

This model is called SETAR because $x_{t-d} = y_{t-d}$, so the AR term itself is the threshold variable. The theoretically infinite number of regimes is limited to 2 or 3, such as high, medium, and low, so this model can be applied in a simplified form, similarly to Eq. (2.15):

$$y_t = \begin{cases} \mu_L + \rho_{L,1} y_{t-1} + \cdots + \rho_{L,pL} y_{t-pL} + \varepsilon_t & \text{if} \quad \theta_L \le x_{t-d} \\ \mu_M + \rho_{M,1} y_{t-1} + \cdots + \rho_{M,pM} y_{t-pM} + \varepsilon_t & \text{if} \quad \theta_L < x_{t-d} \le \theta_H \\ \mu_H + \rho_{H,1} y_{t-1} + \cdots + \rho_{H,pH} y_{t-pH} + \varepsilon_t & \text{if} \quad \theta_H < x_{t-d} \end{cases} \tag{2.15}$$

LSTAR is an acronym for *the Logistic Smoothing Threshold AutoRegression* model, which includes a logistic smoothing AR(p) structure:

$$\begin{aligned} y_t = & \left(\mu_L + \phi_{L,1} y_{t-1} + \cdots + \phi_{L,p} y_{t-p} + \varepsilon_t\right) \cdot G\left(z_t, m, \gamma\right) \\ & + \left(\mu_H + \phi_{H,1} y_{t-1} + \cdots + \phi_{H,p} y_{t-p} + \varepsilon_t\right) \cdot \left(1 - G\left(z_t, m, \gamma\right)\right) \end{aligned}$$

where $G(\cdot)$ for the logistic function, z_t is the threshold variable and x_t plus the delay order (delay order) m, so $z_t = x_{t-m}$.

Because SETAR and LSTAR divide the sample into several regimes using the smooth transition function, if there is too little data, there are insufficient observations and it is impossible to determine the interval transition. If the data is non-stationary, unit root warnings and prediction failures are likely. To compare with ARIMA, US core inflation is used as an example.

The subsequent section uses the same symbol y for US core inflation rate. The script is listed below:

2.2.2 RLab. The SETAR Model for US Core Inflation

Code Chunk. Inflation Level

```
1. out.setar=setar(in_sample, m=4, mL=3, mH=2)
2. fcst1.setar=predict(out.setar,
                              n.ahead=length(out_sample))
3. accuracy(fcst1.setar,x=out_sample)
```

In terms of the script, Line 1 is the estimation that fits the in_sample to setar, within which **m=4** is the AR terms (from AR(0) to AR(3)) for the threshold variable (embedding dimension) and the estimation result is stored as the object, `out.setar`.

The setar() function declares many parameters, as follows:

```
setar(x, m, d=1, steps=d, series, nthresh=1,
mL, mM, mH, thDelay=0, mTh, thVar, th,
trace=FALSE, nested=FALSE, include = c(
"const", "trend","none", "both"),
common=c("none", "include","lags", "both"),
model=c("TAR", "MTAR"), ML=seq_len(mL),
MM=seq_len(mM), MH=seq_len(mH), trim=0.15,
type=c("level", "diff", "ADF"),
restriction=c("none",
"OuterSymAll","OuterSymTh"))
```

d is the time lag control period, for which built-in is 1.

nthresh=1 is the number of threshold values. The default is 1, so two-state intervals (regimes: low & high) are generated. If more are needed, this is changed, but the maximum is also 2, which corresponds to three-state intervals.

mL, mM, mH represent the number of AR orders for the low/middle/high regimes. If these are not declared, the value is m. If each interval requires its own AR order, then it will be declared one by one.

thVar is the threshold variable. If no exogenous variables are declared, the default is its own m-term lagged AR(m-1) linear combination.

The screen print of Fig. 2.12 shows the estimation results.

For the output in Fig. 2.12, the threshold value is 5.698, which separates two regimes: low and high. As specified, the low regime is AR(3) and the high regime is AR(1). The bottom line of Fig. 2.12 shows that the observation value for the low regime (i.e., $Z(t) < 5.698$) is 84.29% and the observation value for the high regime (i.e., $Z(t) \geq 5.698$) is 15.71%.

Using `plot(out.setar)`, Fig. 2.13 visualizes the threshold value and the regimes that are identified are shown in Fig. 2.14. The high regime is marked in red and mainly occupies the middle part of the time series.

```
> out1.setar

Non linear autoregressive model

SETAR model ( 2 regimes)
Coefficients:
Low regime:
     const.L        phiL.1        phiL.2        phiL.3
 0.025718104  1.111895700  0.006064931 -0.124798246

High regime:
  const.H    phiH.1
0.0897870 0.9870131

Threshold:
-Variable: Z(t) = + (1) X(t)+ (0)X(t-1)+ (0)X(t-2)+ (0)X(t-3)
-Value: 5.698
Proportion of points in low regime: 84.29%       High regime: 15.71%
```

Fig. 2.12 Estimation output for object `out.setar`

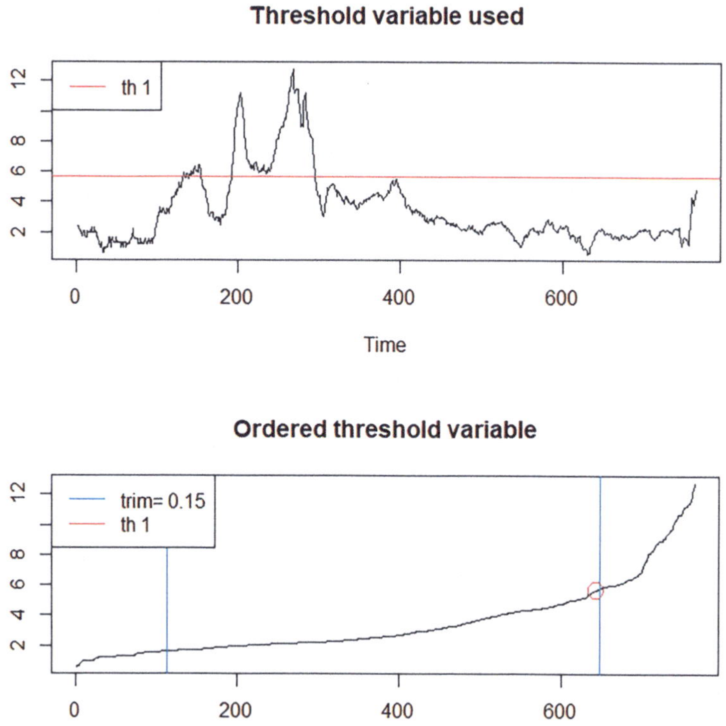

Fig. 2.13 Time segmentation for threshold variables. Both graphs plot the threshold value (5.698). After 1980, inflation enters lower regime, so after 1980, the inflation rate is less than 5.7%, and after 2000, it is about 2.5%

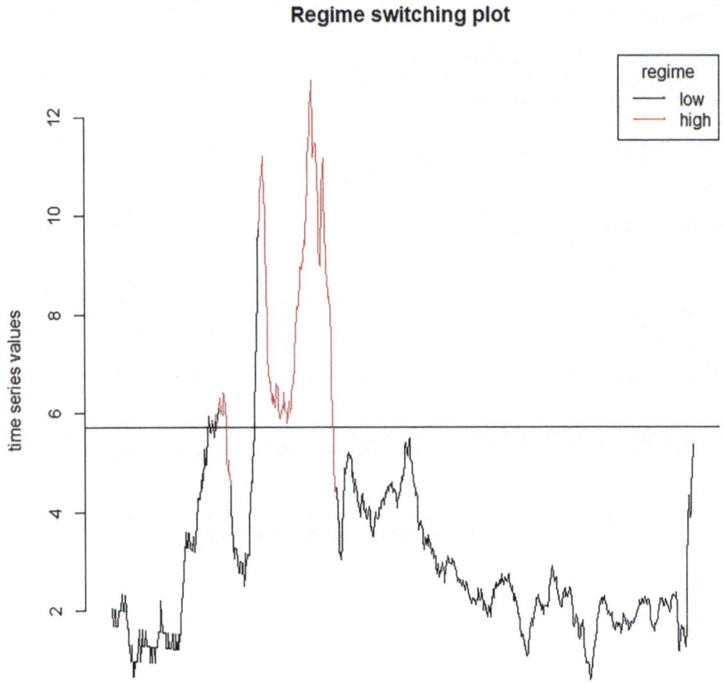

Fig. 2.14 Regimes marked on the time series and separated by the horizontal threshold value (5.698). The higher portion defines the higher regime, and the bottom portion is the lower regime. Colors can be viewed by executing code for this chapter. The plot shows that both regimes feature a structural break: After 1980, the inflation rate is in the low regime so after 1980, the inflation rate is less than 5.7%, and after 2000, it is around 2.5%

Line 3 calculates the forecasting performance and compares it with that for ARIMA. The nonlinear model is not efficient. Figure 2.15 plots the forecasting results for SETAR, and the poor performance is compared with that for Fig. 2.15.

What follows is forecasting practice using the LSTAR model.

2.2.3 RLab. The LSTAR Model for US Core Inflation

Code Chunk. Inflation Level

```
1.  out.lstar=lstar(in_sample,m=4,mL=3, mH=2)
2.  fcst1.lstar=predict(out.lstar,n.ahead =
length(out_sample))
3.  round(accuracy(fcst1.lstar,x=out_sample)
```

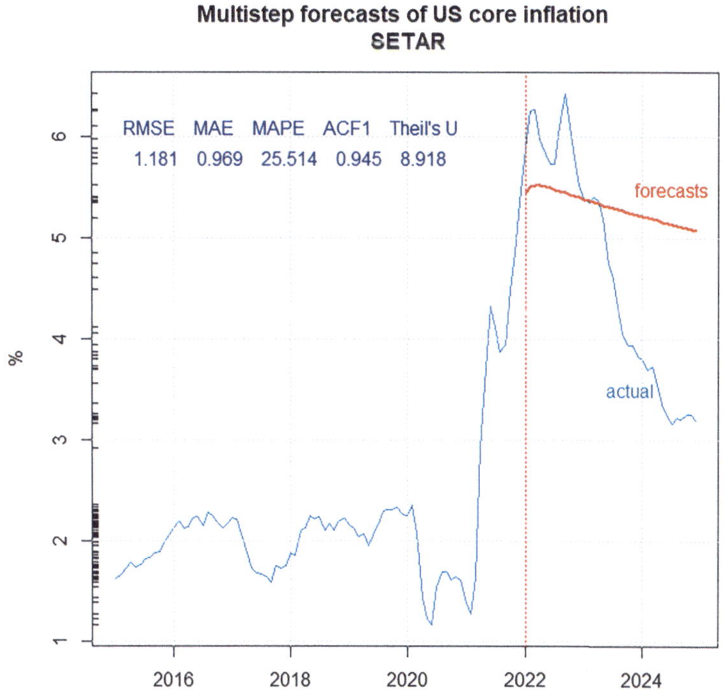

Fig. 2.15 SETAR multistep forecasts of inflation. The results show that the out-of-sample multistep forecasts are inaccurate. The RMSE (1.181) is much greater than the value for the ARIMA counterpart

The accuracy measures are shown in Fig. 2.16. The poor performance is similar to that for SETAR.

Similarly, inflation difference is fitted to LSTAR and multistep forecasts are generated, as detailed in the Chapter Code. It is not discussed here.

R Exercises

1. Retrieve the daily Dow Jones Index and take log of trading volume. Adjust the in-sample periods to 3 years and the out-of-sample to 20 days. Use the SETAR and LSTAR models to generate an out-of-sample multistep forecast for the log volume.
2. Calculate the daily range volatility for the Dow Jones Index using the formula $\frac{(\ln \max - \ln \min)^2}{4 \ln 2}$ and then predict this range volatility as Exercise 1.

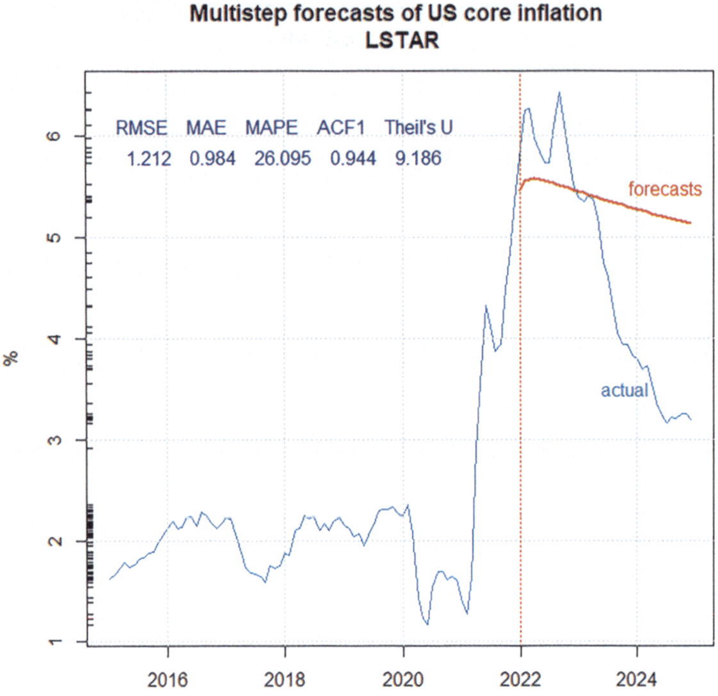

Fig. 2.16 LSTAR multistep forecasts for inflation. The out-of-sample multistep forecasts are as inaccurate as those for SETAR. The RMSE (1.212) is much greater than the value for the ARIMA counterpart

2.3 BATS (Box-Cox transform, ARMA, Trend, Seasonality)

2.3.1 Model Explanations

As noted in Sect. 2.1, the inclusion of seasonal components affects the estimation and prediction performance for the model. Traditionally, dummy variables are used to capture the fixed structure of the season. However, if the seasonal structure features unknown time series properties, it cannot be captured using dummy variables with fixed time labels, and this is termed complex seasonality. If the original data is monthly data, then there are two layers of seasonality: month and quarter. If it is weekly data, there are three layers of seasonality. Taylor (2003) extended the linear version of Holt-Winters and proposed a model with two seasonal components, as expressed by Eq. (2.16):

$$
\begin{aligned}
y_t &= L_{t-1} + b_{t-1} + s_t^{(1)} + s_t^{(2)} + e_t, \qquad i = 1,2\cdots T \\
L_t &= L_{t-1} + b_{t-1} + \alpha \cdot e_t \\
b_t &= b_{t-1} + \beta \cdot e_t \\
s_t^{(1)} &= s_{t-m_1}^{(1)} + \gamma_1 \cdot e_t \\
s_t^{(2)} &= s_{t-m_2}^{(2)} + \gamma_2 \cdot e_t
\end{aligned}
\tag{2.16}
$$

where m_1 and m_2 are seasonal periods and e_t is a white noise variable that represents the disturbance term. L_{t-1} and b_{t-1}, respectively, represent the intercept and the trend of y_t at time t, s_t is the seasonal component, and the coefficients α, β, γ_1, and γ_2 are smoothing parameters. $\left\{s_{1-m_1}^{(1)},\ldots,s_0^{(1)}\right\}$and $\left\{s_{1-m_1}^{(2)},\ldots,s_0^{(2)}\right\}$ represent the initial state variables. The last two formulae for Eq. (2.16) assume that the seasonal component has two types, which is similar to the state-space method.

However, this model has several drawbacks in that it features unstable nonlinearity and it does not recognize complex seasonal cycles. Therefore, De Livera et al. (2011) advocated the use of the BATS method to predict time series that feature complex seasonal trends. This method is named after its four components: Box-Cox transformation, ARMA, trend (Trend), and seasonality (Seasonality). The Box-Cox transformation is a widely used transformation method because it stabilizes the variance of the time series. BATS is defined using Eq. (2.17):

$$
y_t^{(\lambda)} = L_{t-1} + \phi b_{t-1} + \sum_{i=1}^{T} s_{t-m_i}^{(i)} + e_t \tag{2.17}
$$

$$
\begin{aligned}
y_t^{(\lambda)} &= \begin{cases} \dfrac{y_t^{\lambda} - 1}{\lambda}, \lambda \neq 0 \\ \log(y_t), \lambda = 0 \end{cases} \\
L_t &= L_{t-1} + \phi b_{t-1} + \alpha \cdot e_t \\
b_t &= (1-\phi) b_{t-1} + \phi b_{t-1} + \beta \cdot e_t \\
s_t^{(i)} &= s_{t-m_i}^{(i)} + \gamma_i \cdot e_t
\end{aligned}
$$

where e_t follows the ARMA(p, q) process.

This model is only suited to low-frequency time series. The previous RLab is continued using the US core inflation.

2.3.2 *RLab. The BATS Model for US Core Inflation*

Code Chunk

```
1.  out1.bats = bats(in_sample,
                     use.box.cox = NULL,
                     use.trend = NULL,
                     use.damped.trend = NULL)
2.  fcst1.bats =forecast(out1.bats,
                     h=length(out_sample))$mean
3.  accuracy(fcst1.bats,x=out_sample)
```

The declarations inside the bats() function are very useful. Readers should examine all the declarable items and practice them.

`use.box.cox` = TRUE means that a Box-Cox transformation should be performed on the data.

`use.damped.trend` = TRUE means that a damping parameter should be added to the trend. If = NULL, both situations with and without the damping parameter are tested and the best result is selected according to AIC.

The forecasting accuracy is shown in Fig. 2.17. The out-of-sample forecasting values diverge from the actual values significantly.

Finally, the pros and cons of the BATS model are briefly described:

Advantages

1. The Box-Cox transformation can be used for nonlinear data and ensures that the variance of the transformed data is close to a constant.
2. The ARMA structure of the residuals solves most sequence-related problems.
3. There is no sensitivity to the initial values.
4. Point predictions and interval predictions are generated.
5. The overall performance of BATS is better than that for simple State-Space models.

Disadvantages

1. The IID assumption for the residuals may not hold.
2. Explanatory variables cannot be added and it cannot be expanded into ARMAX, like ARMA.
3. The time structure of multiple seasons must be nested layer by layer.
4. For high-frequency data seasonality, a large number of parameters must be estimated.

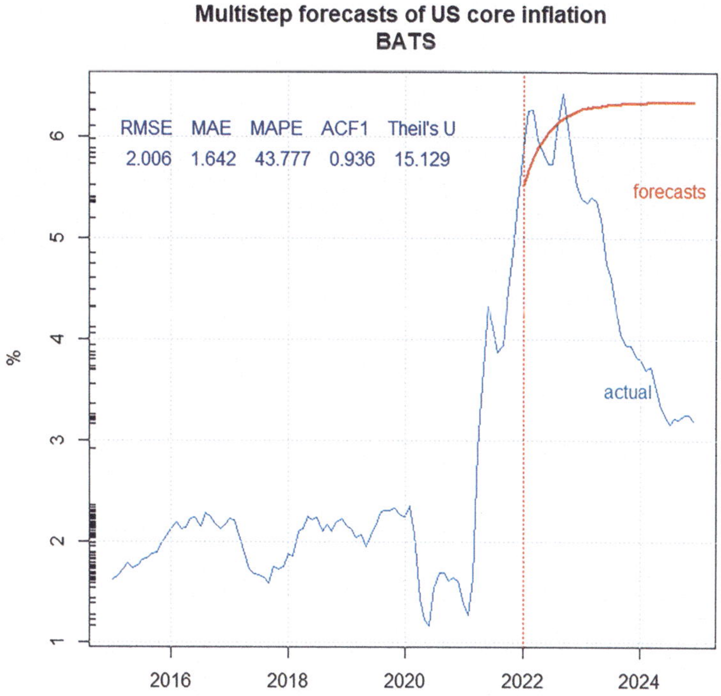

Fig. 2.17 BATS multistep forecasts for inflation. The out-of-sample multistep forecasts are worse than those for previous models, and the RMSE (2.006) is much greater

2.4 BAGGED (Bootstrap AGGregation) Model

2.4.1 Model Explanations

For univariate time series modeling, exponential smoothing is a forecasting method that uses the history of the time series process and most of the methods for calculating forecasts use the concept of exponential smoothing. The concept of exponential smoothing is that the forecasting value uses a weighted average technique: the farther the observation value is from the present, the smaller is the weight, and the closer the observation value, the greater the weight. The exponent is significant because as the observation value occurs farther in time, the weight decreases exponentially (exponential decreasing).

Similarly to BATS, Bergmeir et al. (2016) proposed a technical approach for bootstrap enhancement (bootstrap aggregation) of exponential smoothing, which significantly increases forecasting performance so it is known by the acronym BAGGED. BATS uses the Box-Cox transformation according to STL decomposition and divides the time series into three parts: trend, seasonality, and residuals. BAGGED does the same.

ETS has two common names: one refers to Error, Trend, and Seasonal and the other is ExponenTial Smoothing. For details about ETS, please refer to Chapter 7 of Hyndman and Athanasopoulos (2018).

Bagged is an enhancement method so it enhances two models: ETS and ARIMA. Therefore, in subsequent RLab practice, ARIMA (*p, d, q*) uses the automatic optimal order algorithm (`auto.arima`). In RLab, BAGGED.ETS is used to indicate bootstrap aggregation-enhanced ETS, and BAGGED.auto.arima indicates bootstrap aggregation-enhanced auto.arima. The case implementation uses the previous section's U.S. monthly inflation.

2.4.2 RLab. The BAGGED Models for US Core Inflation

The BAGGED procedures generate 100 models, each of which has three information criteria: AIC, AICc, and BIC. The smallest model in terms of any one of the indicators is omitted and used to calculate the static prediction. With respect to ETS and ARIMA, there are two Code Chunks below with the same structure. Line 1 is the main function `baggedModel`, which generates 100 models, and Line 2 extracts the model that has the smallest `ic` value and computes multistep forecasts using this model.

Code Chunk. Bagged ETS model

```
1.  bagged.ets = baggedModel(in_sample, fn=ets)
2.  ID1= which.min(sapply(bagged.ets$models, AIC))
3.  M1=bagged.ets$models[[ID1]]
4.  fcst1.ets=forecast(M1, h=length(out_sample))$mean
```

Code Chunk. Bagged auto.arima model

```
1.  bagged.arima = baggedModel(in_sample, fn=auto.arima)
2.  ID2=which.min(sapply(bagged.arima$models, AIC))
3.  M2=bagged.arima$models[[ID2]]
4.  fcst1.arima =forecast(M2, h=length(out_sample))$mean
```

Figure 2.18 plots both results. `bagged.ETS` performs as BATS, and `bagged.arima` features similar performance to `auto.arima` in Sect. 2.1.

In terms of static forecasting, it is not possible to directly put the object in the first line. This program currently does not support static bootstrapping forecasting. The R exercise question 2 in this section shows how this is done.

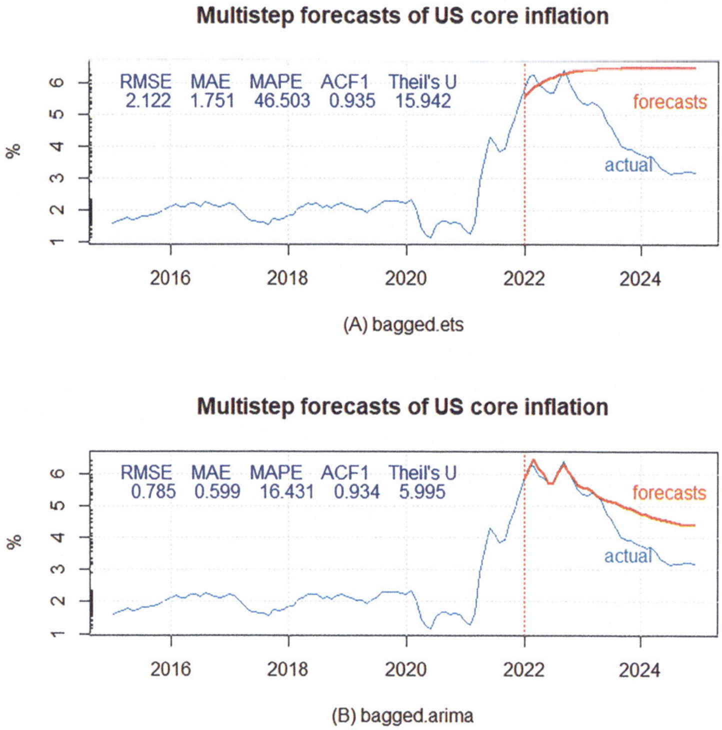

Fig. 2.18 BAGGED multistep forecasts for inflation. The out-of-sample multistep forecasts for ETS are as bad as those for BATS, where RMSE (2.122) is too big. The bagged ARIMA is as good as ARIMA

Next is the use of the BAGGED enhancement model of `auto.arima`. The entire structure is the same as that for ets but the first line of **fn=auto.arima** differs.

R Exercises

1. In the second line of the ETS RLab, select and omit out the model with the smallest BIC and determine the effect on the forecast performance.
2. Calculate the forecast results for all 100 models and evaluate the performance to determine whether the best model is also the best in BIC/AIC/AICc?
   Hint: use

```
lapply(bagged.ets$models,
function(x) forecast(x,ahead=length(out_sample))$mean)
```

to calculate the estimation results for 100 bootstrapping models

2.5 GAMs

2.5.1 Model Explanations

Nonparametric regression does not need to satisfy the assumptions of linear correlation. It can flexibly explore the complex relationship between data, but when the number of explanatory variables in the model increase, the estimation variance for the model increases. It is also difficult to explain the relationship between independent variables and dependent variables in a nonparametric regression that is estimated using kernel and smooth splines. Stone (1985) proposed additive models, within which each additive term uses a single smoothing function for estimation and explains how the dependent variable changes with the independent variable in each additive term. Hastie and Tibshirani (1990) expanded the application of additive models and proposed generalized additive models.

Generalized additive models (GAMs) are modified using linear smoothing models and are nonparametric extensions of generalized linear models (GLM). They retain the basic framework of a GLM and do not feature as many assumptions as other linear models (such as normal assumption or variance homoscedasticity). They are better suited to nonlinear models than other models and are suited to overly complex nonlinear relationships between many variables. Spatial data (such as meteorological, air pollution data) is usually nonlinear, which limits the application of a GLM, so it can be solved using a GAM.

Depending on the regression setting, GAMs have the general form:

$$E\left(y_t | x_{1t} | x_{2t} |, \ldots x_{pt}\right) = \mu + f_1\left(x_{1t}\right) + f_2\left(x_{2t}\right) \cdots + f_p\left(x_{pt}\right) \tag{2.18}$$

where f_j is an unspecified nonparametric smoothing function that is represented as a weighted sum of basis functions:

$$f\left(x_t\right) = \sum_{j=1}^{m} s_j\left(x_t\right)\beta_j \tag{2.19}$$

where s_j is the weight that is applied to the j-th basis function. To estimate a model using this basis expansion, the least squares regression is used. The smooth function is represented by different types of basis functions. To expand the basis function to build a model, the result is calculated using the least squares method. GAMs are configured using a scatter point smoothing method for each function, such as a cubic spline or a kernel smoothing function, so GAMs have sufficient flexibility to include any combination of linear and nonlinear components.

In terms of time series applications, Wood and Augustin (2002) extended the penalty regression splines approach by taking advantage of generalized spline smoothness (GSS). This section uses the method of Wood and Augustin (2002) to adopt a nonparametric generalized additive autoregression form:

$$y_{t+s} = \mu + \sum_{j=1}^{m} s_j \left(y_{t-(j+1)d} \right) \tag{2.20}$$

where s_j are nonparametric univariate functions of lagged time series values. In R, these are represented by cubic regression splines. s_j are estimated together with their level of smoothing using routines in the **mgcv** package. The subsequent RLab continues to use the US core inflation.

2.5.2 *RLab. The GAM Model for Core Inflation*

The `tsDyn::aar()` function is used, where `aar` is the abbreviation for nonparametric generalized additive autoregression. `mgcv::gam()` is used to perform the estimation, but this function must handle the data structure of the time series internally after the estimation, before computing multistep forecasts. `tsDyn::aar()` is much easier to implement.

Code Chunk

```
1.  out1.gam=tsDyn::aar(in_sample, m=1, d=3)
2.  fcst1.gam=predict(out1.gam,
                        n.ahead=length(out_sample))
3.  accuracy(fcst1.gam, x=out_sample)
```

Estimation results and multistep forecasts are sensitive to the specification of `m` and `d`. Line 1 specifies `m=1` and `d=3` to obtain the best accuracy measure but this is not sufficiently accurate, as Fig. 2.19 illustrates.

2.6 ARFIMA—A Fractionally Differenced ARMA Process

2.6.1 *Model Explanation*

An ARFIMA model is like ARIMA but the non-integer differencing order *d* is different. FI denotes a fractional difference order. In other work, it is also known as long memory, which describes a time series process that is close to a unit root process, but not non-stationary.

The estimation process is similar to that in the Sect. 2.1 ARIMA, so the technically theoretical elements are saved here as the step-by-step algorithm:

For an ARFIMA(2, *d*, 0) model, the fractional differencing parameter *d* is estimated using `fracdiff` and the data is fractionally differenced using the estimated value for *d*.

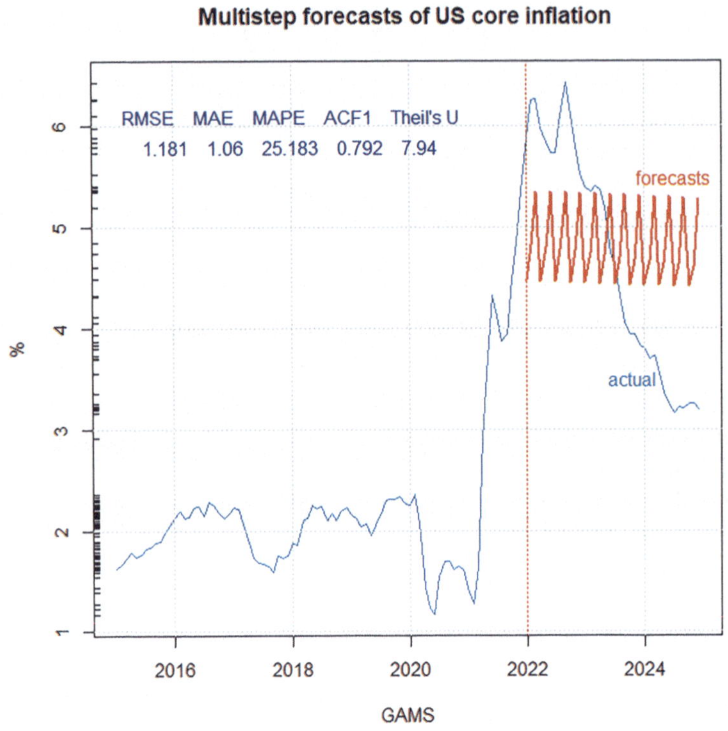

Fig. 2.19 GAMS multistep forecasts for inflation, which show a cyclical pattern that is far from the actual values

An ARMA(p, q) model is then selected for the resulting time series using `auto.arima`.

The full ARFIMA(p, d, q) model is finally re-estimated using `fracdiff`. If `estim == "mle"`, and the ARMA coefficients are refined using `Arima`.

In R, an ARFIMA(p, d, q) model is selected and estimated automatically using the Hyndman and Khandakar (2008) algorithm to select p and q and the Haslett and Raftery (1989) algorithm is used to estimate the parameters, including *d*. This function uses `fracdiff` and `auto.arima` to automatically select and estimate an ARFIMA model.

2.6.2 RLab

The function is offered by package forecast and the code chunk below illustrates an example.

Code Chunk. Multistep Forecasts using ARFIMA

```
out1.arfima=arfima(in_sample,
                     drange = c(0, 0.95),
                     estim = c("mle", "ls")[1],
                     model = NULL,
                     lambda = "auto",
                     biasadj = T)
fcast.arfima = forecast(out1.arfima,
                             h=length(out_sample))$mean
```

Figure 2.20 plots the results, according to accuracy measures. A fractional differenced ARMA does not outperform the `auto.arima` in Sect. 2.1 but it features forecasting performance that is superior to the performance of several nonlinear models.

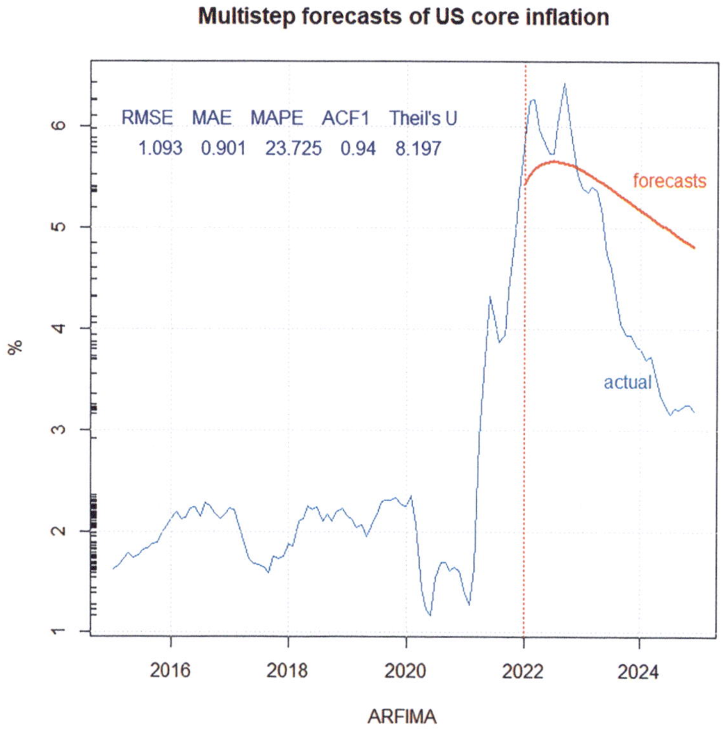

Fig. 2.20 ARFIMA multistep forecasts for inflation. In terms of the RMSE/MAE, a fractional differenced ARMA does not give a better forecast than `auto.arima`. ACF1 = 0.94 also implies higher serial correlation

2.7 Additional Topics

2.7.1 *Model Averaging*

For multiple models, each model's prediction performance is different. From the model's perspective, there is model uncertainty, so predictions are generated by combining the best predictions for multiple models (combination forecast). Multi-model prediction averaging improves accuracy because the prediction that is generated by the model is either overestimated or underestimated relative to the true value, so the average is used. In terms of econometric methods, Bates and Granger (1969) first proposed the concept of combination forecasting and showed that when there are multiple models or other predictions that are generated using multiple settings for a model, combining them using a weighted average method improves the final prediction. Combination forecasting requires the calculation of weights, so econometrics uses a model average method that assigns a weighted average to the predictions that are generated using multiple settings for the prediction model. Hansen (2007) proposed a Mallows model averaging (MMA) method and Hansen and Racine (2012) proposed a Jackknife Model Averaging (JMA) cross-validation method and provided a more theoretical basis for the method of prediction averaging. Recently, two Taiwanese econometricians, Liao and Tsay (2020), extended the concept to VAR prediction.

Hansen's (2007) MMA is quite similar to the solution for quadratic regularization. The data observation has T items so to combine the M predictions for the model, the best weight is determined by solving the following mathematical program:

$$\arg\min_{w} C_n\left(w\right) = w'\Omega w + 2\sigma^2 K'w$$

In this objective function, w is the weight vector, Ω is the M × M covariance matrix of prediction errors, K is the M × 1 positive vector that represents the number of parameters for each model, and σ^2 is the variance for the entire complete model.

The JMA method of Hansen and Racine (2012) is similar to the calculation of influence statistics for the Stability test (as is the diagnosis of outliers, such as Cook distance). The observation values are removed one by one to generate one-by-one prediction results and the JMA solves the following objective function:

$$\arg\min_{w} C_n\left(w\right) = w'\Omega w + 2\sigma^2 K'w$$

$\tilde{Y}$ is a T × M matrix, each column of which represents a model0specific prediction, so there are M columns. Observations of each column represent the t-th predictive value that the m-th model predicts after removing the t-th observation value. This is the same type of cross-validation that is used for conventional K-Fold cross-validation.

To calculate weights, the theoretically based MMA and a simple equal weight averaging method, such as the method of Aiolfi et al. (2011), can be used. The equal weight method is very commonly used because if there are future uncertainties, equal weighs give the best result and there is no need to spend time calculating the best prediction weight from the sample information. DeMiguel et al. (2009), Tu and Zhou (2011), and Pflug et al. (2012) came to similar conclusions. Pflug et al. (2012) theoretically proved this point using simulation methods. When the model uncertainty for the future is high, the equal-weighted portfolio is the best out-of-sample investment strategy.

In machine learning literature, this is also known as the committee approach (see Bishop, 2006, (Ch.14) and Hastie et al. (2009, Ch.8)). This book uses these methods later. For this case, neither MMA nor JMA is suitable because both of these weight calculations use a model, such as linear regression with at most K explanatory variables, so the prediction estimate is taken from the sub-variables in the K variables, which is very similar to the random forest method. However, if there are 10 different models, this method is not suitable because it lacks the variance σ^2 of the full variable models and the elements of K may all be the same.

2.7.2 AveW

The window prediction averaging method (AveW: Average Windows Forecasts) of Pesaran and Pick (2011) is the original version of AveW, and the principle is shown in Fig. 2.21: Estimation denotes the in-sample window, which estimates generates parameters, and test defines the out-of-sample window for multistep forecasts. The average of many multistep forecasts is the AveW.

The text by Meese and Rogoff (1983) shows that random walk is the best forecasting option for exchange rates. Random walk is expressed as $y_t = y_{t-1} + e_t$, so the best forecast of tomorrow's exchange rate is the exchange rate today. If t denotes today, then the AveW version of random walk is expressed as:

$$\hat{y}_{t+1} = mean\left(y_t + y_{t-1} \cdots + y_{t-m}\right)$$

This is similar to the simple average for several long-horizon forecasts or a moving average. Observations are sequentially omitted, the model is then re-estimated, and multistep forecasts are generated. AveW is model-specific and is similar in execution to the 100 models of bagged arima that were discussed previously. Pesaran and Pick (2011) determined the effect of window size on prediction performance for a single model. I practice, several results are averaged to improve forecasts.

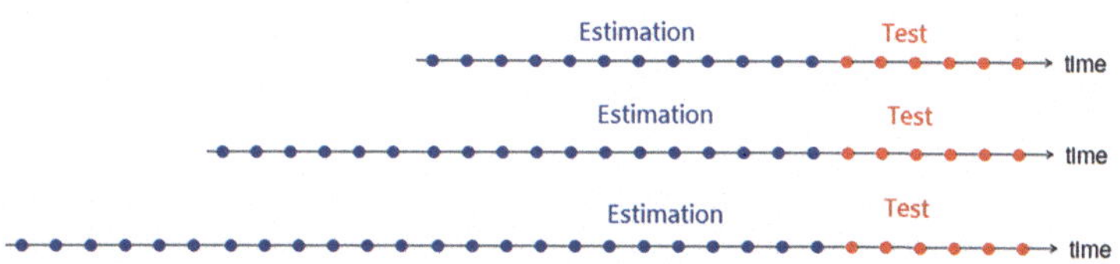

Fig. 2.21 An outline of AveW: AveW is a framework that uses a moving window and a single model so all forecasts are averaged out of moving windows

Model averaging is not a panacea, and the average of many poor forecasts does not result in a good prediction. If one model mostly over-predicts the true values and the other mostly under-predicts the true values, then the average may be close to the true values. Models must generate forecasts in two opposite directions to make the average meaningful.

Interested readers might try to average the forecasts using SETAR, LSTAR, and BATS and to determine whether the average using these models improves predictive performance.

2.7.3 Diebold–Mariano Test for Predictive Accuracy

The Diebold–Mariano statistic (1995) tests for the equality of forecast accuracy for two forecasting errors under general assumptions. The function also corrects for the autocorrelation that multi-period forecast errors usually exhibit. An efficient h-period forecast has forecast errors for MA(h-1) processes. The Diebold–Mariano test uses a Newey–West type estimator to determine the sample variance of the loss differential to account for this concern. The package `forecast` features the function `dm.test` to implement this strategy. To compare the results for SETAR and auto.arima, their respective prediction errors are calculated as:

```
e.arima= Fcst1-out_sample
e.setar= fcst1.setar-out_sample
```

The Diebold–Mariano test is specified as:

```
dm.test(e.setar, e.arima, alternative="greater")
```

where the test defaults Forecast horizon = 1 and Loss function power = 2. Forecast horizon = 1 indicates h = 1 and `Loss function power = 2` implies that the loss function that is used is RMSE, or MAE if the `Loss function power =1`.

The null is E2=E1, and the `alternative="greater"` implies that if the null is rejected by E1 (the latter), E1 outperforms E2 (the former). The output is:

```
> dm.test(e.setar, e.arima, alternative="greater")

      Diebold-Mariano Test

data:  e.setar-e.arima

DM = 6.0217, p-value = 3.614x10^-05

alternative hypothesis: greater
```

It is clear that `auto.arima` outperforms SETAR for multistep forecasts, which is shown by the plots. The winner for the Diebold–Marino test must not justify to conclude a good forecaster. Visualization is a useful tool in this respect.

2.7.4 Onestep Ahead Forecast and Moving Windows

Unlike multistep forecasts, which recursively compute out-of-sample forecasts over h steps, a onestep forecast is estimated using a predictive regression, such as Eq. (1.2). To predict the future for 36 months using a onestep ahead forecast, a multiple onestep forecast is used so a onestep forecast is executed 36 times by moving the window, as shown by Fig. 2.22.

The R demo uses data CPIx.csv. This file also contains two more variables: interest rate spread and unemployment rate. The interest rate spread is the difference between the 3-month Treasury Bill rate and the Federal Fund rate, and unemployment is determined using the Phillips curve. To execute a onestep forecast, this text uses the method of Stock and Watson (1998) for inflation modeling and forecasting.

A predictive regression is fitted as:

$$\text{inf}_t = \alpha + \beta_1 \times \text{spread}_{t-1} + \beta_2 \times \text{unrate}_{t-1} \tag{2.21}$$

Hence, a 3-variable data frame INF is created as follows:

Code Chunk. Data manipulation

```
library(forecast)
tmp=read.csv("./data/CPIx.csv")
head(tmp)
y0=ts(tmp, start=c(1957,1),freq=12)
y=diff(log(y0[,2]),12)*100 #Core Inflation
dat0= na.omit(data.frame(y,y0[-seq(12),-c(1:2)]))
dat=embed(as.matrix(dat0),2)
INF=cbind(dat[,1],dat[,5:6])
colnames(INF)=c("inf","spreadL1","unrateL1")
 INF =ts(INF, end=c(2024,12),freq=12)
```

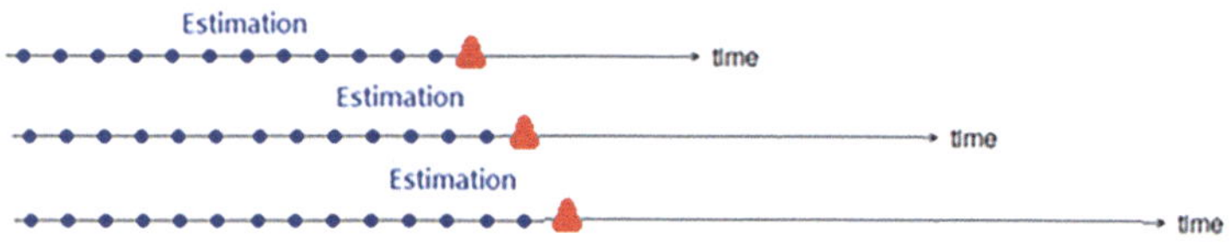

Fig. 2.22 Moving-window onestep forecasts. A onestep prediction can be repeated many times using a moving window

The following script creates an object `J` that sets the time intervals for onestep loop estimation and onestep prediction. Eq. (2.21) is fitted and because the model generated by `auto.arima` is ARIMA(2,1,2)(2,0,0), an ARIMA(2,1,0)(2,0,0) and exogeneous variables are specified. This is named ARIMA(2,1,0)(2,0,0)+X. The MA(2) terms are omitted because the importance of the MA terms is to be determined.

Line 1 defines a symbol `J`, and Line 3 is the main loop that structures the moving-window forecasts.

Code Chunk. Moving-window onestep forecast

```
  1.  J=data.frame(Year=c(2021,rep(2022:2024,each=12)),
                     Month=c(12, rep(1:12,3)))
  2.  fcst1.arx=NULL
  3.  for (i in (2:(nrow(J)-1))) {
    mw_in=window(INF,end=c(J[i-1,1],J[i-1,2]))
    mw_out=window(INF,start=c(J[i,1],J[i,2]))
    out0=Arima(mw_in[,1],order = c(2,1,0),
                     seasonal=c(2,0,0), xreg = mw_in[,-1])
    if (i==36) {
         fcst1.arx=c(fcst1.arx,
                        forecast(out0,h=2,
                                    xreg =mw_out[,-1])$mean)
    } else {
      fcst1.arx=c(fcst1.arx,
                     forecast(out0,h=1,
                               xreg =mw_out[,-1])$mean[1])
    }
}
```

Figure 2.23a plots the results for ARIMA(2,1,0)(2,0,0), and the accuracy measures are the out-of-sample comparison. Onestep ahead predictions do not perform well so a onestep ahead forecast using a moving-window auto.arimax: `auto.arima with xreg` is also conducted. Interested readers can refer to the chapter code file for detail. The result is plotted as Fig. 2.23b. The moving-window onestep forecast `auto.arima` (2,1,2)(2,0,0)+`X` substantially outperforms the forecast using ARIMA(2,1,0)(2,0,0) +X.

To end this chapter, the last example shows that moving-window onestep ahead prediction gives excellent performance but it is practically infeasible when confronted with the real future. Forecasting is neither a self-exciting laboratory task nor a historical back-testing exercise. Using machine learning, the baseline is multistep forecasting and `auto.arima`. The machine learning model is used in a multistep framework.

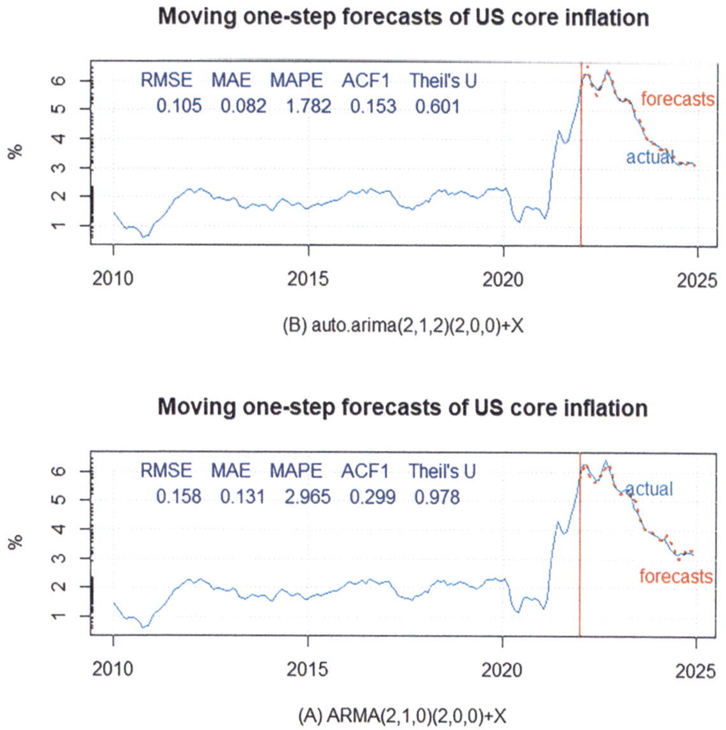

Fig. 2.23 Moving-window onestep forecasts: 2022M1–2024M12. The dotted line is the prediction. The moving onestep is performed by re-estimating the new window and predicting one value ahead. The results show that in terms of the accuracy measures, `auto.arima`(2,1,2)(2,0,0)+X substantially outperforms ARIMA(2,1,0)(2,0,0) +X and that the role of MA(2) is critical

R Exercise

- Conduct a Diebold–Mariano test for Fig. 2.23 and determine whether the missing MA(2) terms are critical for forecasting.

Chapter 3
Forecasting Using Machine Learning Methods

3.1 Regression Tree, Random Forest, and KNN

3.1.1 Model Explanations

A decision tree is a structure under $y = f(X)$. If the explained variable y is discontinuous data of {0, 1}, it is called a classification tree. In econometrics, this is referred to as a limited dependent variable, and it is known as a target variable in data mining. If y is continuous data, it is called a regression tree. Depending on the type of Y, the tree model is divided into two types: a classification tree, where Y is categorical, such as a binary integer {0, 1} or multiple {1, 2, 3, 4}, or a regression tree, where Y is a continuous value.

In data mining, the tree algorithm generates a series of "if-then" logic conditions (if..., then...) such that the X variables explain (predict) the y variable. Decision trees are generally tree-shaped structures that are generated from top to bottom. Each decision or event end (also known as a natural state) can produce two or more different results. This decision structure looks like the branches of a tree. Figure 3.1 explains a simple $y = f(X1, X2)$ classification process, where Ey represents the expected value of y:

X1 is divided into two sections at value 1.7:

1. If $X1 < 1.7$, then $Ey = 2.5$
2. If $X1 \geq 1.7$, then it is divided into two sections:

   2.1. If $X2 < 202.1$, then $Ey = 5.6$
   2.2. If $X2 \geq 202.1$, then $Ey = 1.3$

Each condition for X1 or the combination of X1 and X2 is called a node, or a classification node. Within this condition, many y observations may be circled (growing a tree) and then their average is the expected value. The generation of node values is the optimal result for a decision tree algorithm.

T.-w. Ho, *Time Series Forecasting using Machine Learning*,
https://doi.org/10.1007/978-3-031-97946-0_3

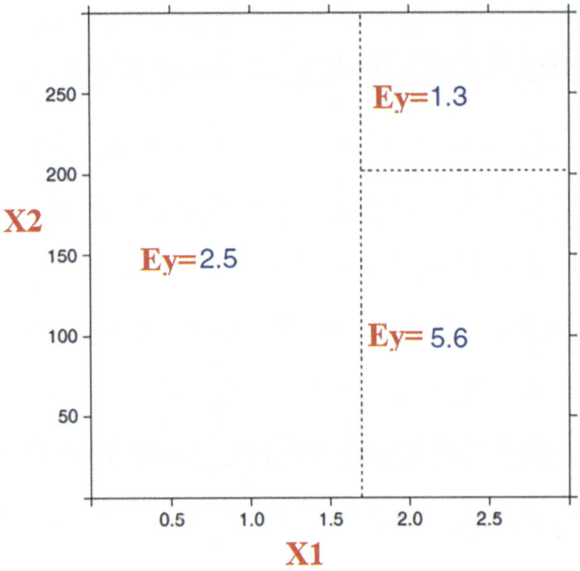

Fig. 3.1 Simple decision tree, explaining the sample splitting structure that separates several regimes in terms of the critical values of covariates

Prediction results that are calculated using the data can be applied to tree-shaped decisions. For example, a bank might have a decision problem: should a credit card be issued to a specific applicant? Using the nodes that are calculated using the data, the decision-maker uses the following "**if**..., **then**..." linguistic structure:

If (1) The applicant's monthly mortgage payment is less than 25% of their income
(2) The mortgage payment is less than 1 month late
(3) The monthly income is more than 50,000 Taiwan dollars
(4) A card is issued

This classification process is drawn as the credit card issuance decision process in Fig. 3.2:

In future, when there are credit card applicants, only the customer's financial data is required to make a decision. If Y is binary or multiple, classification is direct.

In terms of another customer analysis example, a bicycle shop owner has had 1200 customers in the past, 289 of whom bought bicycles. The owner decides to analyze potential buyers based on "age," as shown in Fig. 3.3:

The two previous examples show that the structure decision trees uses sample splitting: The subsamples of target variable and each independent variable are gouped as a tree node. This classification criterion is different to principal components or clusters, which do not use known independent variables to classify specific dependent target variables. Decision trees must have this information. To determine whether a node value of 40 years old is a correct age threshold, entropy and Gini

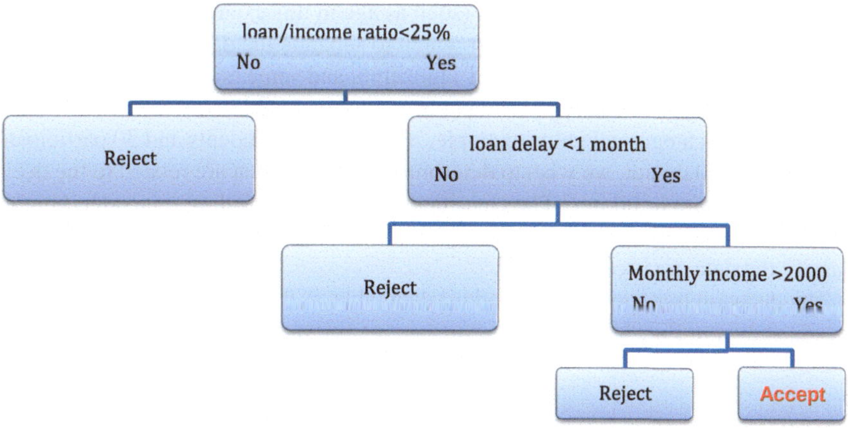

Fig. 3.2 Credit card issuance decision process

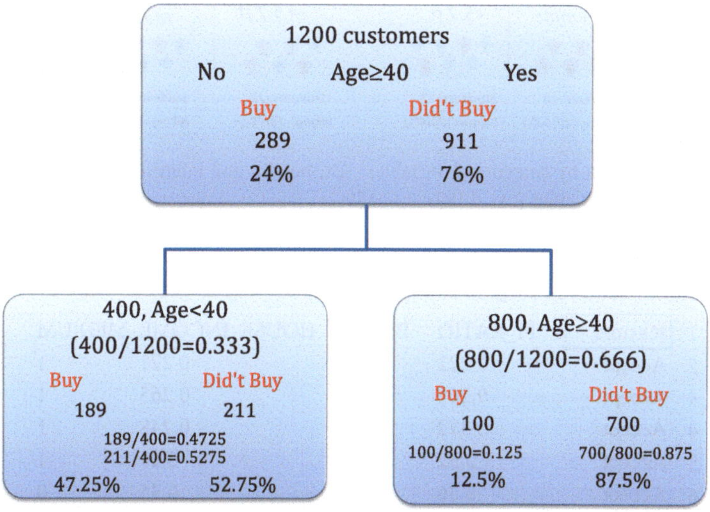

Fig. 3.3 Customer structure for whether to buy a bicycle

numbers are used. The Gini number measures the impurity of the model: the minimal Gini number is the key value. The calculation of the Gini number for the structure in Fig. 3.3 is written as:

Left	Right
$0.333 \times 0.4725 \times (1 - 0.4725)$	$0.666 \times 0.125 \times (1 - 0.125)$

The Gini number is the sum of **left + right = 0.1558**. The smaller the Gini number, the better. Therefore, the algorithm calculates the Gini numbers for various classifications using known variables (such as age in this example) and the smallest number is the best threshold (node) at which to separate the data (grow a new leaf). The Gini number is used by the CART algorithm, and some models use information gain, which calculates entropy. The details are not included in his text.

For another categorical value example, if there are 50 students and 30 participate in biking and 20 do not, we wish to determine the factors that are related to the decision to participate in biking? The classification results according to gender, height, and academic expertise are shown in Fig. 3.4. Using the Gini, the variables that best classify the data are used.

The decision tree model is a standard data table. A data table for mortgage issuance is shown in Fig. 3.5.

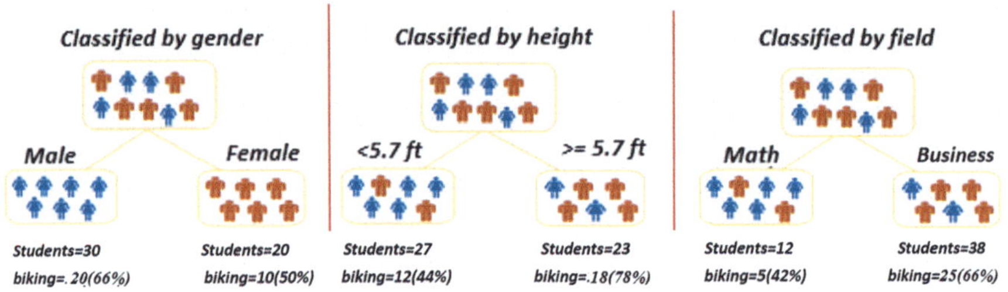

Fig. 3.4 Classification by categorical variables: sex, height, and interest. This figure shows that different classifying variables give different results

	A	B	C	D	E
1	Results	PI_RATIO	BLACK	HOUSE_INCOME	MEDIUM
2	Accept	0.221	0	0.221	1
3	Accept	0.265	0	0.265	1
4	Accept	0.372	0	0.248	1
5	Accept	0.32	0	0.25	1
6	Accept	0.36	0	0.35	0
7	Accept	0.24	0	0.17	0
8	Accept	0.35	0	0.29	0
9	Accept	0.28	0	0.22	1
10	Reject	0.31	0	0.24	1
11	Accept	0.18	0	0.17	0
12	Accept	0.28	0	0.27	0
13	Accept	0.31	0	0.3	0
14	Reject	0.28	0	0.27	0

Fig. 3.5 Loan issuance data. Column A, Results, is the target binary variable, which is decoded by {Accept, Reject} which is the result of a loan application. The choice is binary so it is also known as a classification problem. Many studies predict binary outcomes in terms of the remaining variables

Column A is the result (Results). This column is Y, so it is the target variable or label. The file variable details are:

Results = Reject, application rejected; Results = Accept, application accepted. This is the target variable (Target variable)
PI_RATIO: The ratio of the applicant's monthly amortization to income
BLACK =1: the applicant is black; =0 is not
HOUSE_INCOME: Applicant's monthly house expense-to-income ratio
DENIED_MI =1: Applicant's mortgage insurance is rejected
MEDIUM =1: Loan-to-value ratio is between 0.8 and 0.95
HIGH =1: Loan-to-value ratio ≥ 0.95
LOAN_VALUE: Loan-to-value ratio is the ratio of loan amount to house price
CONSUMER_CREDIT: Applicant's consumer credit score
MORTGAGE_CREDIT: Applicant's mortgage credit score
PUBLIC =1: there is a public bad credit record
SELF =1: self-employed
SINGLE =1: single
URIA: Unemployment rate for the industry in which the applicant works

The explanatory variable X chooses PI_RATIO and BLACK. The parameter values for the decision tree "If-then" algorithm are:

(1) root 1785 216 Accept (0.88 0.12)
(2) PI_RATIO < 0.40995 1592 148 Accept (0.91 0.10) *
(3) PI_RATIO >= 0.40995 193 68 Accept (0.65 0.35)
(6) PI_RATIO < 0.582 171 53 Accept (0.69 0.31)
(12) BLACK < 0.5 130 32 Accept (0.75 0.25) *
(13) BLACK >= 0.5 41 20 Reject (0.49 0.51)
(26) PI_RATIO < 0.435 19 6 Accept (0.68 0.32) *
(27) PI_RATIO >= 0.435 22 7 Reject (0.32 0.68) *
(7) PI_RATIO >= 0.582 22 7 Reject (0.32 0.68) *

This numerical result is drawn as a tree structure in Fig. 3.6.

The original data fits this classification structure to produce a probability prediction, as shown in Fig. 3.7. The rows are the data for the loan applicants, and the decision tree generates the probability for each applicant's binary state: {Accept, Reject}. The state with the higher probability becomes the predicted value, in the far right column.

The predicted values for the binary state {Accept, Reject} and the actual values are listed side by side in Fig. 3.8.

The prediction of binary time series data is discussed in a later chapter. When the algorithmic points for the decision tree are defined, the random forest emerges by growing many trees randomly.

The data table in Fig. 3.5 is used. The dependent variable RESULTS is the target, and 13 explained variables are the top layer **All Data**. The random forest randomly selects a subset of variables from the 13 explanatory variables (**Random subset** layer) and then executes a decision tree algorithm (**tree** layer) to obtain a prediction

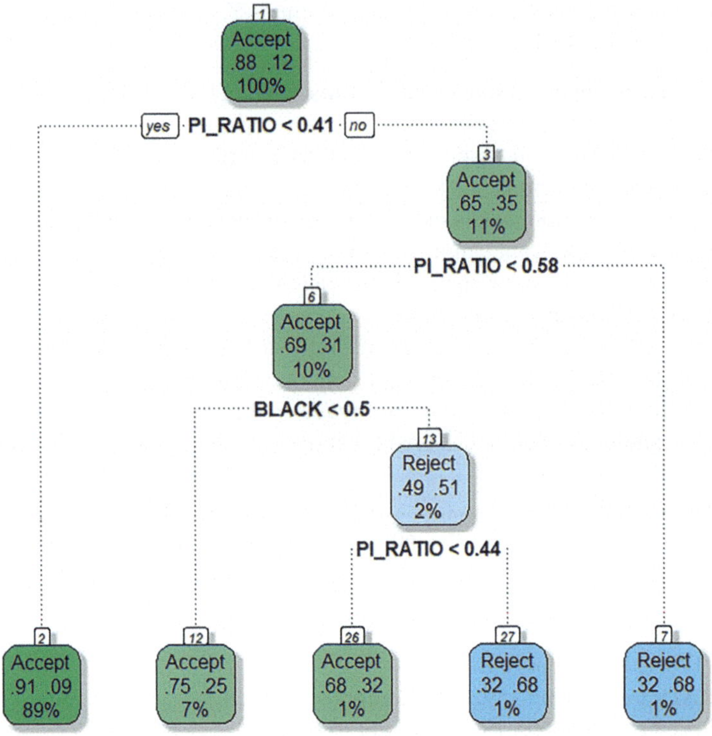

Fig. 3.6 Tree structure for decision tree classification of data in Fig. 3.5. The number at the bottom of each square is the percentage of observations within that square, and the upper title labels the character of the larger percentage

Fig. 3.7 Data table for the prediction values that are generated by a decision tree

A	B	C	D
Accept	Reject	Prediction	
0.3182	0.6818	Reject	
0.9070	0.0930	Accept	
0.9070	0.0930	Accept	
0.9070	0.0930	Accept	
0.7538	0.2462	Accept	
0.3182	0.6818	Reject	
0.9070	0.0930	Accept	
0.6842	0.3158	Accept	
0.6842	0.3158	Accept	
0.9070	0.0930	Accept	
0.9070	0.0930	Accept	
0.9070	0.0930	Accept	
0.9070	0.0930	Accept	
0.3182	0.6818	Reject	
0.9070	0.0930	Accept	
0.7538	0.2462	Accept	
0.9070	0.0930	Accept	

Fig. 3.8 Comparison of predicted results and actual results

A	B
Result_Raw	Result Prediction
Accept	Accept
Accept	Reject
Reject	Accept
Accept	Accept
Accept	Accept
Reject	Accept
Accept	Accept

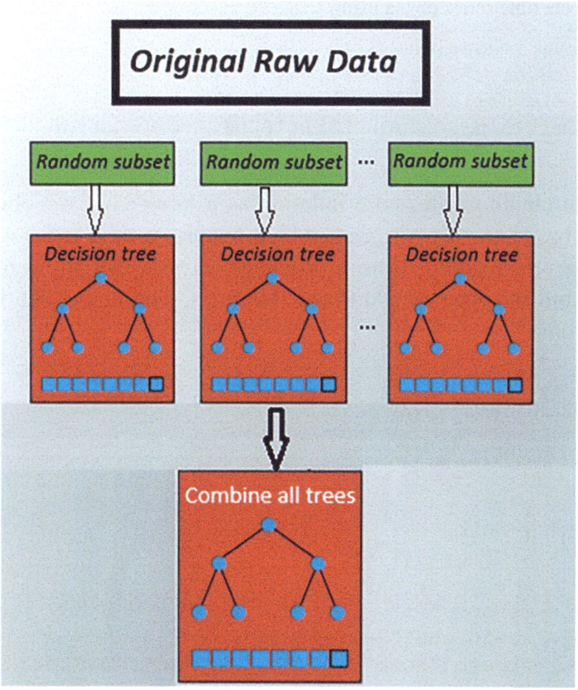

Fig. 3.9 Random forest visualization showing how a forest is created using many trees from random subsets

result for Results (as shown in Fig. 3.5). If $C_5^{13} = 1287$, 500 tree prediction results are generated. The random forest selects variables randomly and then combines the results. To predict the first result using 500 decision trees, if there are 300 Accepts and 200 Rejects, so the prediction result for the random forest is Accept. The random forest is a method of reinforcement, which gives robust results that conform to data attributes (Fig. 3.9). In Figure 3.9, it explains how a forest is created using many trees from random subsets.

The subsequent RLab uses the American core inflation data in Chap. 2 and then uses AR(6) to demonstrate random forest multistep forecasts for R.

3.1.2 RLab. Multistep Forecasts Using Random Forest

In terms of machine learning, the time series object used is `timeSeries`, and the estimation and calculation of multistep forecasts use `iForecast`. The following code chunks represent the three stages of time series forecasting using random forest: data manipulation, estimation, and forecast calculation.

Code Chunk. Create time series object using `timeSeries`

```
library(timeSeries)
tmp=read.csv("./data/CPI.csv") #core CPI
y0=as.timeSeries(tmp[,2], as.Date(tmp[,1]))
y=na.omit(diff(log(y0),12))*100 #Core Inflation
```

This R code chunk creates core inflation as a `timeSeries` object. The next code chunk is based on `iForecast`, which features a time series wrapper `tts.caret` that uses the machine learning package `caret`. Its options manipulate the AR structure and the type of additional dummies. The next Code Chunk shows these descriptions.

Code Chunk. Estimation using `iForecast`

```
library(iForecast)
output1.rf = tts.caret(
          1.  y=y,
          2.  x=NULL,
          3.  arOrder=c(1:6),
          4.  xregOrder=c(0),
          5.  method="rf",
          6.  train.end="2021-12-01",
          7.  type=c("none", "trend", "season", "both")[1],
          8.  tuneLength =10,
          9.  resampling=c("boot", "cv", "repeatedcv")[1],
          10.  preProcess = "center")
```

The details of main options are given in Help.

Line 1 and Line 2, respectively, define the target (dependent) variable and input (independent) variables. Both variables must be for the full sample and not subset training data. The training data is controlled by the argument `train.end` in Line 6.

Line 3. The `arOrder` defines the specification of the lag structure for the target variable y. This allows either continuous lags c(1, 2, 3,4,5) or discontinuous distributed lags, such as c(3, 6, 9).

Line 4. `xregOrder` defines the specification of the lag structure for the explained variables x. Starting with 0, it allows either continuous lags, such as c(1, 2, 3, 4), or discontinuous distributed lags, such as c(0, 3, 6, 9).

Line 5. `method` defines the train_model_list for `caret`. The method must allow regression. Methods in c(`"rf","svm","gbm","enet","nnet","knn"`) are feasible and common, and c(`"xgbDART", "xgbTree", "xgbLinear", "avNNet"`) are time-consuming.

For binary classification, c(`"glm", "rpart", "lasso", "bridge", "pls", "nb"`) are suitable, where `rpart` and `nb`, respectively, denote a decision tree and Naive Bayes.

The K-Nearest Neighbor (KNN) method defines the argument `method="kknn"`. `"knn"` is used for binary or multivariate dependent variables.

Line 6. `train.end` is a date object, which delimits the in-sample range that is used to estimate the data frame for y and x. For a rolling window, the dates are reset. Using the example in Chap. 2, the in-sample for estimation ends with "2021-12-01".

The remaining lines explained because they are very straightforward.

This chunk estimates the data, and the multistep forecasts are calculated as:

Code Chunk. Multistep forecasts using random forest

```
1.  testData=window(output1.rf$dataused,
                start="2022-01-01",
                end=end(y))
2.  fcst1.rf=iForecast(Model=output1.rf,
                    Type="dynamic",
                    n.ahead = nrow(testData))
3.  fcst1.rf=as.timeSeries(fcst1.rf,
                     as.character(time(testData)))
4.  accuracy(x=as.ts(testData[,1]),
          object=as.ts(fcst1.rf))
```

Using the example in Chap. 2, the out-of-sample for multistep forecasts starts from 2022-01-01, so

Line 1. The out-of-sample testing data is retrieved.

Line 2. Multistep forecasts are calculated using `iForecast()` with the argument `Type="dynamic"` and `n.ahead` is the number of observations for testData, where `n.ahead` is the same as `h` in ARMA dynamic forecasting.

Line 3. The forecasting values are converted into the `timeSeries` object.

Figure 3.10 plots two forecasting results: with and without temporal dummies. In terms of the RMSE and MAE, random forest gives better predictions than models in Chap. 2, except ARIMA.

Random forecast does not generate exactly the same results because the forest grows randomly. To obtain the same results, the `seed` option is defined at the beginning.

Finally, data analysis calculates variable importance using the command:

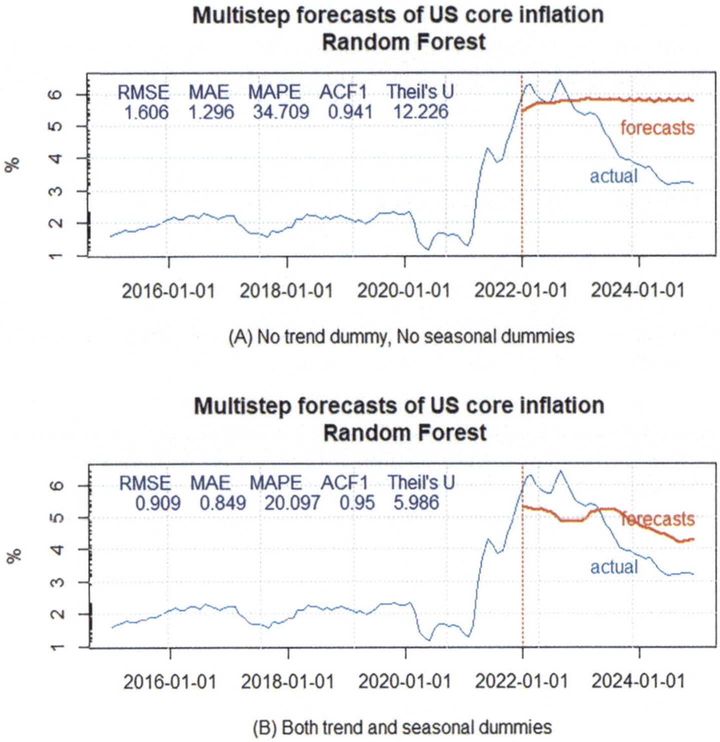

Fig. 3.10 Prediction performance for two types of random forest models: (**A**) No trend dummy and no seasonal dummies; (**B**) with trend dummy and seasonal dummies. It indicates that using a deterministic trend or seasonal dummies improves the prediction accuracy

```
varImp(output2.rf$output)
```

Since `varImp()` is a wrapper, it calls a function that calculates relative importance and, `varImp()` calls a function-specific calculation.

R Exercises

1. Conduct a Diebold–Mariano test to compare the predictive accuracy of Figs. 2.10 and 3.10.

3.2 Simple Artificial Neural Network

When we have many variables, and the data table is like Fig. 3.11. If economic growth is the dependent variable, assuming a linear relation with other covariates may be not proper. Neural network is a way to consider unknown relationship.

	A	B	C	D	E	F	G	H	I	J
1		Economic Growth, %	Export, %	Import, %	IPI, %	TWUSD_AVG, %	TWUSD_EOP, %	INF, %	Long rate, %	Short rate, %
198	2011Q1	6.72	10.54	7.95	11.19	-7.96	7.65	1.39	1.41	0.56
199	2011Q2	4.16	4.01	1.34	2.56	-9.97	-11.24	1.92	1.43	0.69
200	2011Q3	3.28	1.12	-3.98	-1.22	-9.05	-6.61	1.36	1.39	0.82
201	2011Q4	0.89	0.94	-5.13	-11.92	-1.55	-0.91	2.01	1.29	0.86
202	2012Q1	1.09	-0.28	-3.14	-2.12	0.71	0.06	1.25	1.28	0.80
203	2012Q2	1.04	-1.28	-2.51	-3.88	2.59	3.66	1.75	1.24	0.79
204	2012Q3	2.18	3.44	2.82	4.98	2.26	-0.46	2.91	1.18	0.81
205	2012Q4	4.48	5.50	4.64	5.15	-3.54	-3.88	1.60	1.14	0.79
206	2013Q1	1.50	3.65	4.93	-0.32	-0.59	0.85	1.35	1.22	0.77
207	2013Q2	2.68	4.00	2.31	4.64	0.99	0.47	0.59	1.30	0.71
208	2013Q3	1.95	1.83	1.19	2.13	0.28	0.59	0.84	1.63	0.69
209	2013Q4	3.73	3.41	4.69	7.05	1.42	2.36	0.34	1.68	0.64
210	2014Q1	4.69	3.76	2.34	3.19	2.72	2.15	1.59	1.62	0.63
211	2014Q2	4.92	4.87	4.18	6.94	0.70	-0.15	1.62	1.54	0.62
212	2014Q3	4.90	7.81	9.00	9.98	0.38	1.24	0.70	1.64	0.62
213	2014Q4	4.38	7.24	6.88	8.71	4.17	5.35	0.60	1.62	0.65
214	2015Q1	4.80	6.48	2.17	7.52	3.97	3.49	-0.61	1.56	0.65
215	2015Q2	1.89	-0.12	2.56	0.21	2.70	3.50	-0.57	1.54	0.62
216	2015Q3	-0.28	-1.95	-0.03	-5.93	6.88	8.68	0.29	1.30	0.56
217	2015Q4	-0.20	-2.13	0.71	-7.46	6.13	4.83	0.13	1.15	0.41
218	2016Q1	-0.09	-5.29	-2.92	-4.56	5.42	4.13	1.99	0.90	0.39
219	2016Q2	1.69	-2.21	-3.36	1.57	4.59	4.05	0.90	0.82	0.34
220	2016Q3	3.00	0.18	-0.35	7.78	-1.50	-4.37	0.34	0.68	0.33
221	2016Q4	3.92	3.16	2.36	9.54	-3.30	-3.06	1.68	1.07	0.39
222	2017Q1	3.24	4.32	3.59	8.20	-6.99	-6.92	0.18	1.14	0.44
223	2017Q2	2.64	2.24	2.09	5.13	-6.93	-6.82	1.00	1.06	0.39
224	2017Q3	3.61	7.66	2.64	5.00	-4.66	-4.33	0.49	1.04	0.41
225	2017Q4	3.71	3.73	-1.44	4.09	-5.37	-6.55	1.21	1.01	0.43
226	2018Q1	3.23	1.92	1.47	5.70	-5.89	-4.82	1.57	1.02	0.42
227	2018Q2	3.40	3.81	1.90	0.02	-1.62	-0.61	1.39	1.00	0.45
228	2018Q3	2.47	0.29	2.72	1.65	1.31	1.99	1.70	0.85	0.53
229	2018Q4	1.97	-2.67	-0.56	-1.06	2.44	2.77	-0.06	0.91	0.57
230	2019Q1	1.84	0.88	1.02	-10.01	5.03	5.46	0.56	0.82	0.58

Fig. 3.11 A table displaying quarterly economic data from 2011 to 2019. Columns include "Economic Growth, %," "Export, %," "Import, %", "IPI, %," "TWUSD_AVG, %," "TWUSD_EOP, %," "INF, %," "Long rate, %," and "Short rate, %." Each row represents a specific quarter, with data showing trends in economic growth, trade, inflation, and interest rates over time. The table provides a comprehensive overview of economic indicators for the specified period

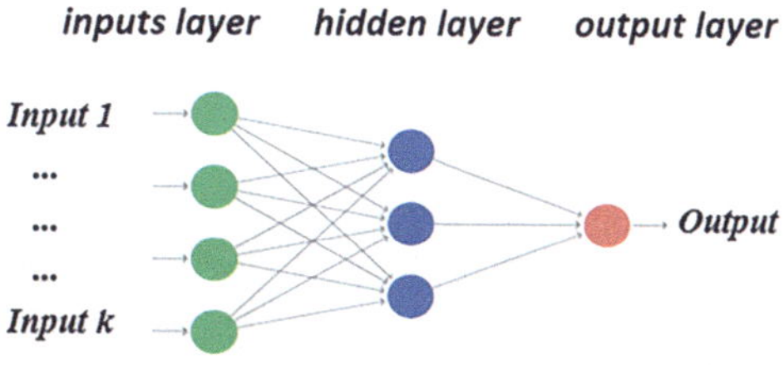

Fig. 3.12 Neural network model architecture. This graph assumes *k* input variables, and the hidden layer consists of three functional variables

3.2.1 Model Explanations

The neural network method includes nonlinear settings for hidden layers and size nodes (hereinafter referred to as NNETAR). This method is a feedforward neural network and uses lagged values of a time series as inputs, as shown in Fig. 3.12. For details of this method, please refer to Chapter 11 of Hyndman and Athanasopoulos (2018).

As shown in Fig. 3.12, the input layer of the neural network is the original data and the hidden layer is regarded as a nonlinear combination of the original data. This linear combination in the hidden layer can be imagined as a series of composite functions (composite functions)

$$x \to f(x) \to g(f(x)) \to h(g(f(x))) \to \cdots \to y$$

Functions *f*, *g*, *h* are generated by algorithms using point-to-point type of data to simulate the operation of the neural system. The generation of the entire cognitive network is complex. In the algorithm for a neural network, the function is generated by a series of neural node (neural nodes) cognitive (perception) systems, which is also known as a layer function.

Figure 3.13 shows the similarities and differences between neurons and decision trees. The decision tree algorithm uses an explanatory variable Xi as a node and then classifies the target variable step by step, so the same Xi can appear multiple times.

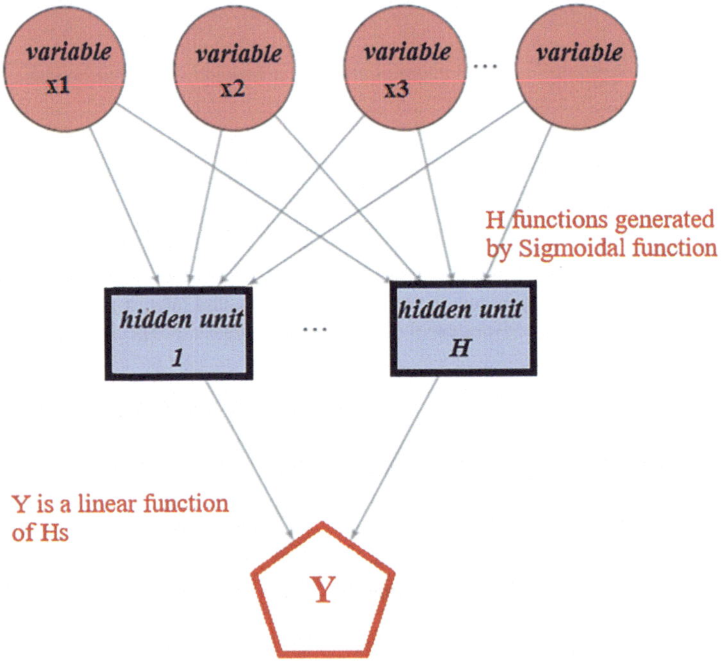

Fig. 3.13 The relationship between multiple neural levels; each layer is a combination of independent variables that are sampled with replacement. Flow chart illustrating a neural network model. The top layer consists of variables(predictors) labeled as ×1, ×2, ×3, and more, represented by red circles. These are connected to a middle layer of hidden units, shown as blue rectangles labeled "Hidden Unit1" and "Hidden Unit H," with a note indicating that H functions are generated by a sigmoidal function. The hidden units connect to a final green circle labeled "Y," with a note stating that Y is a linear function of the hidden units. Arrows indicate the flow of information from predictors to hidden units to the output Y

The neural structure is the same, but the neural node or neuron is a nonlinear combination of multiple explanatory variables in a new function, which is called a hidden function or a hidden layer. For example, for the 20 input variables of $\{x_1, x_2,\ldots x_{20}\}$, the node variable H may be a unknown function:

$$H_A = f^A\left(x_1, x_5, x_7, x_{11}\right)$$
$$H_B = f^B\left(x_7, x_{13}, x_{17}, x_{20}\right)$$
$$\ldots$$
$$\ldots$$

Layer by layer, the final result fits the best Y. The neural algorithm simulates the operation of brain neurons and regards input variables as external stimuli, transmits them layer by layer, and finally produces a brain signal (Y). A neural algorithm generates a suitable number of synthetic functions (layers, layers). Figure 3.13 shows this relationship.

A neural network method has three characteristics:

1. In a multi-layer feedforward network, each layer of nodes receives inputs from the previous layer
2. Linear combinations are used to place inputs at each node
3. The result is modified using a nonlinear function before output

The input to a hidden neuron (neuron) with a linear combination is:

$$z_j = b_j + \sum_{i=1}^{4} w_{i,j} y_i \tag{3.1}$$

The neural network activation function is modified by a nonlinear function (sigmoid):

$$s(z) = \frac{1}{1+e^{-z}} \tag{3.2}$$

The use of nonlinear functions reduces extreme input values, so the network is not affected by outliers. The characteristics of artificial neural networks are:

1. The weights start with random values and are updated using observations.
2. Predictions include randomness, so the network is trained from different random starting points and the result is averaged.
3. The number of hidden layers and the number of nodes in each hidden layer must be decided before the process begins.

NNETAR uses a Box-Cox transformation and a value of $\lambda = 0.5$ to ensure that the residuals are approximately homoscedastic. The model is written in the following form:

$$y_i = f\left(y_{t-i}\right) + \varepsilon_t$$

where f represents a neural network with a single layer and four hidden nodes. It is assumed that the residual term ε_t is homoscedastic (and can be normally distributed) because ε_t generates a random value and repeatedly simulates future sample paths, whether it is generated from a normal distribution or resampled from historical values.

The simple neural network model that was proposed by Franses and Van Dijk (2000) is also used. This is a nonlinear AR with a hidden layer and a linear output:

$$y_{t+s} = \beta_0 + \sum_{j=1}^{D} \beta_j \cdot h\left(\delta_{0j} + \sum_{i=1}^{m} \delta_{ij} y_{t-(i-1)d} \right) \quad (3.3)$$

where D is the delay order and m is the embedding dimension, which is similar to SETAR. The model is referred to as NNET. In addition to standard neural networks, machine learning often uses a Recurrent Neural Network (RNN) and Long Short-Term Memory (LSTM). This type of RNN is widely used for time series prediction.

In `iForecast`, the `tts.caret()` function is used and the argument `method="nnet"` is defined.

3.3 Support Vector Machines

3.3.1 Model Explanations

SVM is termed a shallow machine learning method that is similar to the role of linear regression in mountainous statistical methods. The concept of SVM was proposed by Vapnik (2000).[1] Unlike the empirical risk minimization (ERM) that is used by statistical learning methods, which minimizes training data, SVM embodies the principle of structural risk minimization (SRM), which minimizes the upper limit of the generalization error. Therefore, a SVM is less susceptible to outliers and fits and predicts nonlinear data, so it has more applications.

A SVM is used for data sequence modeling. The method is originated from classification problems. Similar to linear programming, it determines a hyperplane that perfectly cuts data points to classify sample data, but it can also be applied to regression and time series forecasting. Support Vector Regression (SVR) uses the concept of a SVM classification model but linear regression uses the least squares method, which minimizes the sum of squared errors (SSE) to determine the estimated value for the regression coefficients so the parameter estimates are easily affected by an observation that deviates from the central tendency. SVR determines a curve (or hyperplane) that fits the sample data for a predetermined prediction error and a tolerance for outliers. SVR can adjust the tolerance for outliers and can achieve nonlinear fitting so it allows more flexible data fitting and prediction than linear regression.

---

[1] The first edition was published in 1995.

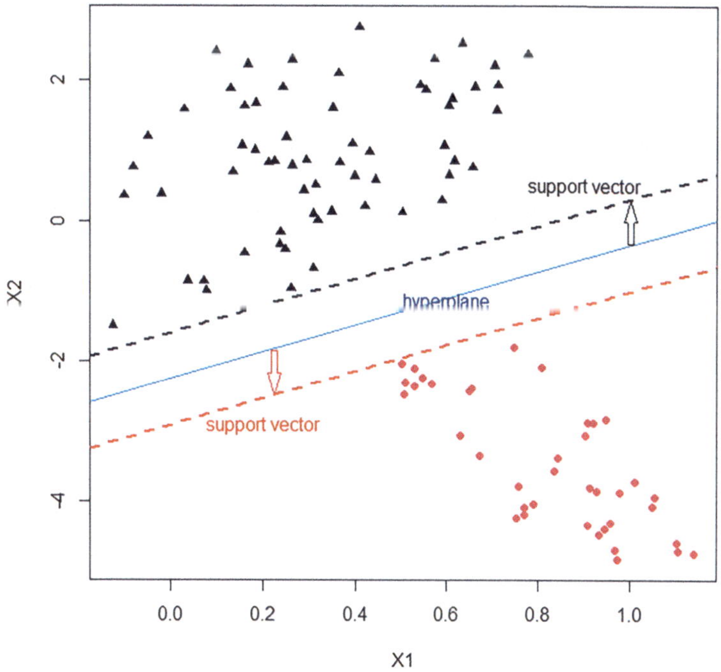

Fig. 3.14 SVC classification diagram. Data points are divided into two groups: black triangles and red circles. A blue line labeled "hyperplane" separates the groups. Two dashed lines, one black and one red, indicate "support vectors." Arrows point to the support vectors, highlighting their role in defining the margin

Starting with SVC, diagrams and mathematical models are used to explain SVM (or SVR). As shown in Fig. 3.14, there are n observations, and SVC selects a small subset as the vector that supports the linear hyperplane.

The subset of SVC is derived using the following optimization function:

$$\max_{\beta_0,\beta_1,\ldots\beta_p} M$$

$$s.t.\sum_{j=1}^{p}\beta_j^2 = 1$$

$$y_i\left(\beta_0 + \beta_1 x_{i1} + \beta_2 x_{i2} + \cdots + \beta_p x_{ip}\right) \geq M\left(1-\varepsilon_i\right), \forall i = 1,\cdots n$$

$$\varepsilon_i \geq 0, \sum_{i=1}^{n}\varepsilon_i \leq C$$

$$\forall y_i \in \{-1,1\} : \text{associated class labels}$$

where y_i is the classification label.

If $y_i = 1$, the observation is blue (left triangle) and if $y_i = -1$, the observation is red (right triangle). This is a binary variable because the linear combination of the hyperplane is 0. If the observation is above this hyperplane (left triangle), $\beta_0 + \beta_1 x_1 + \beta_2 x_2$ is a positive number so multiplying all observations $y_i = 1$ gives a positive result, which satisfies the condition $M > 0$. If the observation is below this hyperplane (right triangle), $\beta_0 + \beta_1 x_1 + \beta_2 x_2$ is negative and gives a positive result when multiplied by $y_i = -1$, which also satisfies the condition $M > 0$. The SVC algorithm solves the parameters in terms of the β values that maximize the value of M.

When this linear classification algorithm encounters observations that are no longer linearly separable or nonlinear, the linear hyperplane does not fully classify the observations, as shown in Fig. 3.15a:

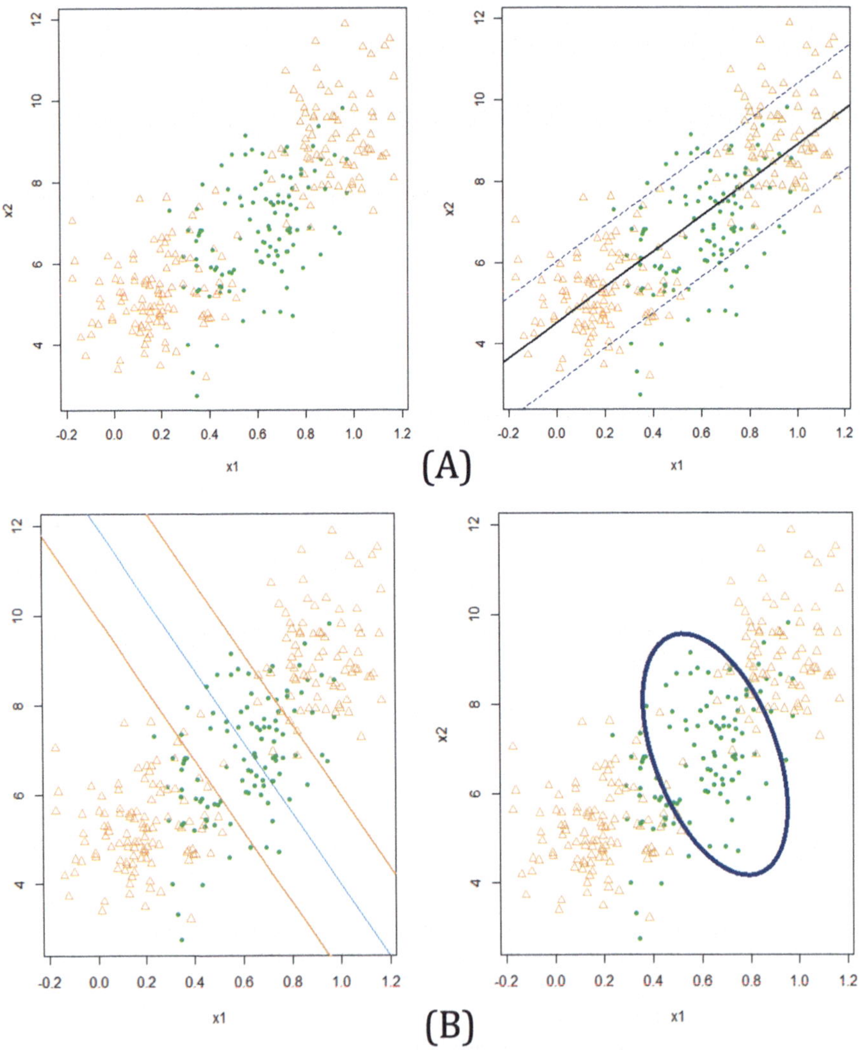

Fig. 3.15 Principle of SVM classification: (**a**) Schematic diagram showing SVM unable to classify observations and (**b**) classification effect of SVM using kernel

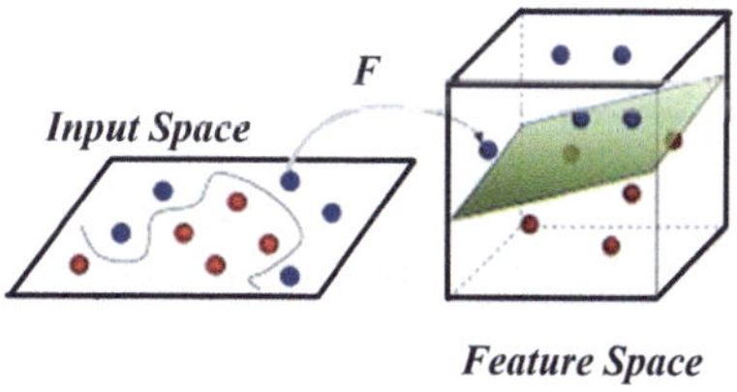

Fig. 3.16 Schematic diagram of a kernel projection function: projection from 2D to 3D contrasts differences and improves classification

SVM uses a kernel function, which is developed from SVC, as shown in Fig. 3.15b. A kernel function projects data from low dimensions to high dimensions (hyperplane) and e data that cannot be classified in two dimensions is then classified, as shown in Fig. 3.16.

In terms of mathematical programming, a SVM solves an optimization function as follows:

$$\max_{\beta_0,\beta_1,\ldots\beta_p} M$$

$$s.t.\sum_{j=1}^{p}\beta_j$$

$$y_i\left(\beta_0+\sum_{j=1}^{p}\beta_{j1}x_{ij}+\sum_{j=1}^{p}\beta_{j2}x_{ij}^2\right)\geq M\left(1-\varepsilon_i\right),\forall i=1,\cdots n$$

$$\varepsilon_i\geq 0,\sum_{i=1}^{n}\varepsilon_i\leq C,\sum_{j=1}^{p}\sum_{k=1}^{2}\beta_{jk}^2=1$$

where C is the cost function.

A SVM uses a different classifier:

$$f\left(x\right)=\beta_0+\sum_{i\in S}\alpha_i x,x_i=\beta_0+\sum_{i\in S}Kx,x_i$$

Linear kernel: $K\left\langle x_i,x_j\right\rangle=x_i^{'}x_j$

There are two types of nonlinear kernels:

1. A d order polynomial "kernel": $K\left(x_i,x_j\right)=\left(r+\gamma\cdot x_i^{'}x_j\right)^d$, $d > 0$, generally $r = 0$ and γ is a constant ($=1/n$).
2. Radial kernel: $K\left(x_i,x_j\right)=e^{\gamma\cdot x_i^{'}x_j}$. This is the most widely used function.

These equations show that a SVM uses the ε loss function and considers the penalty terms. SVM regression coefficients minimize the following "cost and penalty terms" equation:

$$Cost\cdot\sum_{i=1}^{n}L_{\varepsilon}\left(y_i-\hat{y}_i\right)+\sum_{j=1}^{P}\beta_j^2,$$

where $L_\varepsilon(\cdot)$ is the ε-insensitive equation and *Cost r* is the cost penalty, which is used to punish large residuals. The penalty for this formula is represented by the reciprocal of the ridge regression (or Tikhonov regularization) or the weighted decay of neural networks, because the penalty is attached to the residuals, rather than the parameters. The SVM's robust techniques are suited to a statistical regression.

SVM and most machine learning methods resemble a computation task to classify categorical variables but similarly to `lm` to `glm` or a regression tree to a decision tree, SVM classification can be directly applied to continuous data. The concept is similar to the confidence interval that is generated by continuous data in statistics. Statistics constructs confidence intervals according to conditional expectations and standard deviation and then a nominated value (=0.05) is used to categorize the explained variable into two types.

The SVM data structure for this text is basically an AR(p) process that generates multistep forecasts. For more details on SVM, please refer to Hastie et al. (2017, Ch. 12).

There are two popular packages for running a SVM in R: `e1071` and `kernlab`. The function `kernlab` is `ksvm()`:

```
ksvm(x, y,
     kernel ="rbfdot",
     kpar = "automatic",
     C = 1,
     epsilon = 0.1)
```

This `kernel ="rbfdot"`, where `rbfdot` is a radial kernel. In the RLab section `ksvm` is used indirectly via the `caret` package and then a definition is made to the effect method="`svmRadial`", which is simpler. Interested readers can study the `kernlab` package.

3.3.2 RLab. Multistep Forecasts Using SVM

Code Chunk

```
output1.svm = tts.caret(y=y,
            x=NULL,
            arOrder=c(1:6),
            xregOrder=c(0),
            method="svm",
            train.end="2021-12-01",
            type=c("none", "trend", "season", "both")[1],
            tuneLength =14,
            resampling=c("boot", "cv", "repeatedcv")[1],
            preProcess = "center")
```

`tuneLength =14` is the built-in cost range for the SVM: 2^{-2}, 2^{-1}, 1, 2, 2^2,..., 2^{11}. The forecasting results and accuracy measures are presented in Fig. 3.17. Two cases are presented.

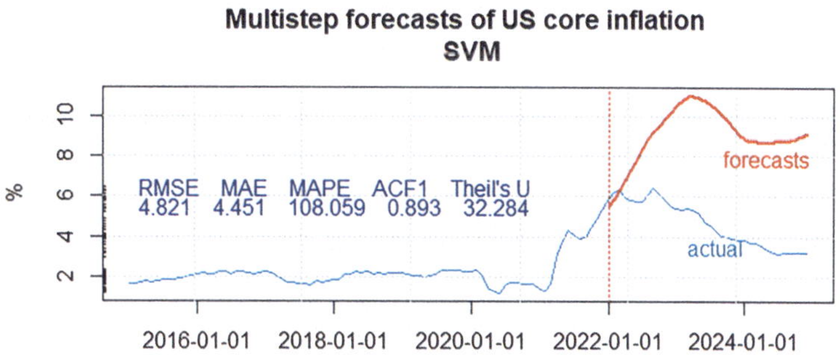

(A) No trend dummy, No seasonal dummies

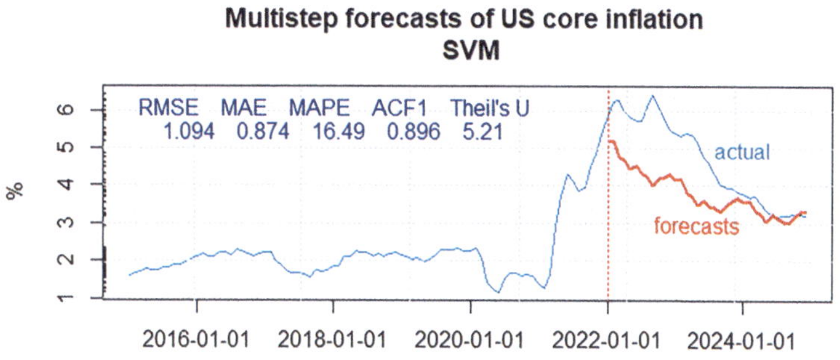

(B) Both trend and seasonal dummies

Fig. 3.17 Out-of-sample multistep forecasts for SVM: similarly to random forest, the inclusion of a deterministic trend and seasonal dummies improves the prediction

The optimization process for a SVM tunes a series of hyperparameters and finally selects the model that has the smallest prediction error.

R Exercise

1. Referring to Fig. 3.17, the average of two models improves the forecasting performance. Confirm this conjecture.

3.4 Gradient Boosting Machine

3.4.1 *Model Explanations*

GBMs (Gradient Boosting Machines) constitute a reinforcement method and were proposed by Friedman (2001).[2] A simple model is used, and the results of new models are added in turn. Although GBMs can be used for any model in a broad sense,

[2] Friedman J. H. (2001) Greedy function approximation: a gradient boosting machine. *Annals of statistics*, 1189–1232.

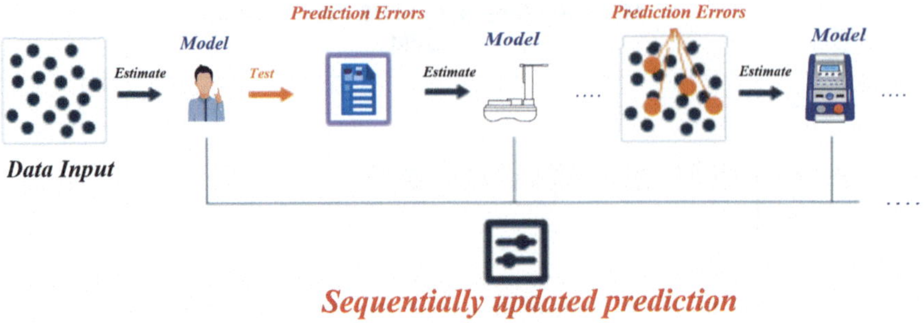

Fig. 3.18 The nature of sequential GBMs is a sequence that continuously updates previous random forest result and improves the residuals

they are generally used for tree-shaped models (either decision trees or regression trees). The difference between a random forest and a GBM is that the reinforcement of the random forest uses the depth of a large forest and GBMs start with a weak, shallow tree. As long as the prediction for the next tree is better, other tree structures are added sequentially, so it is also a sequential reinforcement method (sequential ensemble approach). In recent years, GBMs have repeatedly given the best results in Kaggle's data science competition. Although a GBM uses a shallow tree, with appropriate parameter values (tuning parameters), it produces a powerful committee. The method is similar to carefully screening suitable prediction results and then averaging them.

A GBM is *sequential*. The boosting begins with a simple tree (a small number of nodes) and then generates new trees to reduce prediction errors. New trees are generated to learn the biggest mistakes that are made by the previous tree. The data structure is used to correct the specific rows of data with larger prediction errors, as shown in Fig. 3.18.

A GBM is not a single model. It acts like a data algorithm framework. Its operational logic can be explained step-wise[3]:

Step 1. Fit the data using a (regression) decision tree: $F_1(x) = y$. The prediction errors are $y - F_1(x)$, which is also known as the residual

Step 2. Fit a decision tree to the prediction errors and extract the predicted values: $h_1(x) = y - F_1(x)$

Step 3. Add the predicted values to obtain the new prediction: $F_2(x) = F_1(x) + h_1(x)$

Step 4. Fit a decision tree to the prediction errors for Step 3: $h_2(x) = y - F_2(x)$

Step 5. Add the predicted values to obtain the new prediction: $F_3(x) = F_2(x) + h_2(x)$

Step 6. Continue this process until the stop condition (such as Cross Validation) is achieved.

The final model is a stepwise additive model of κ independent trees:

[3] Boehmke and Greenwell (2020, Chap. 12).

$$f(x) = \sum_{\kappa=1}^{K} f^{\kappa}(x)$$

The six-step algorithm that is described can be illustrated with a simulation, as shown in the regression tree in Fig. 3.19, where the red represents the GBM process, the blue curve is the optimal prediction, and the goal of the sequential process is to minimize the sum of the squared errors (SSE):

Starting from number 0 means there is no tree included, resulting in the sample mean; then stacking begins, an increase in the number of trees corresponds to an increase in the number of new trees in training, until 20, which is almost the theoretical prediction curve.

A simple GBM model features two types of hyperparameters: boosting hyperparameters and tree-specific hyperparameters.

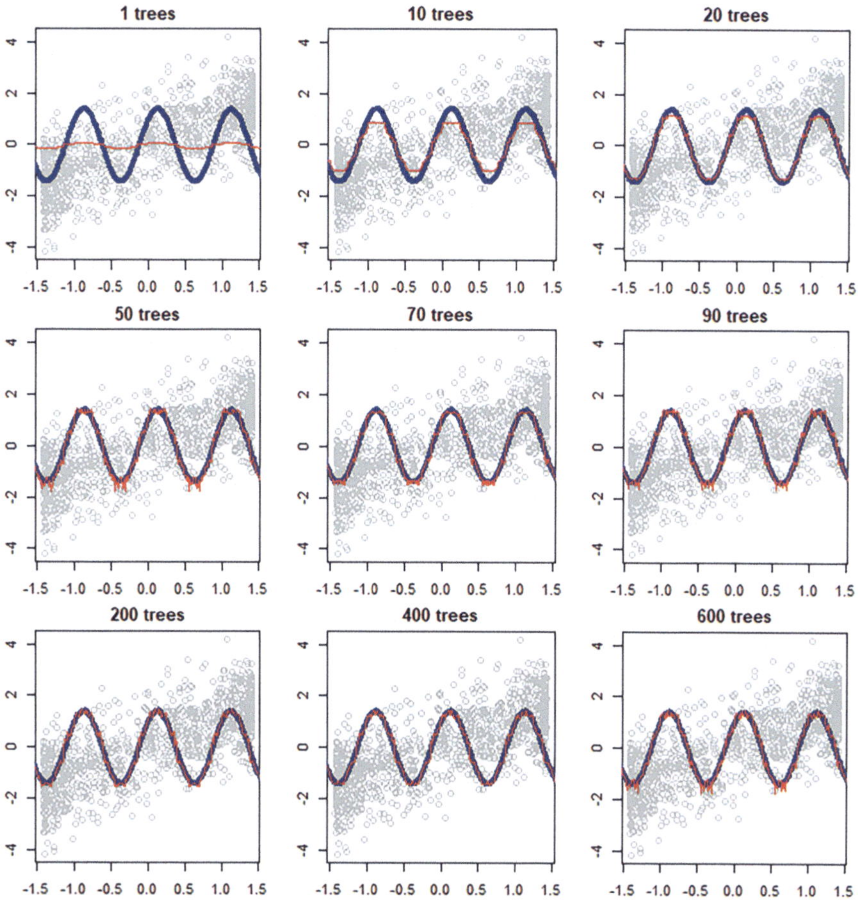

Fig. 3.19 GBM is enhanced by sequential regression trees. Data points are represented as gray circles, while the fitted model is shown as a blue line. As the number of trees increases, the blue line more closely follows the sinusoidal pattern of the data points

There are two boosting hyperparameters:

1. Number of trees: The total number for the entire sequential boosting process. For BAGGing and random forests, the independent results are combined by averaging, which results in overfitting if there are many trees; 20 trees look OK, but the prediction is less accurate if too many trees are averaged, such as 600. In a GBM, each tree revises the prediction errors for the previous tree so as long as the prediction error meets the stopping threshold, it must stop and it is unnecessary to use all the trees.
2. Learning rate: The learning rate determines the contribution of each tree to the final goal and controls the speed at which the algorithm's gradient descends, and gradient descent corresponds to learning. The learning rate is typically n between 0.001 and 0.3. A smaller number increases the robustness of the model to the features of each tree, so it is easy to generalize and convergence is guaranteed, so a smaller number stops the operation and operational risks may merge. The learning rate parameter is also called the shrinkage parameter and the smaller this number, the more accurate is the model, but more trees are required to achieve this goal.

There are also two tree-specific hyperparameters:

1. Tree depth: The typical depth ranges from 3 to 8. A shallow causes the algorithm to jump to a specific classification, which is more efficient in terms of computation, but requires more trees to confirm the overall result. Similarly, deep trees allow the algorithm to classify more unique results, so overfitting is more likely.
2. Minimum number of observations per node: This controls the complexity of each node's classification. One observation per node or all observations per node is not suitable. The typical classification is 5~15 observations.

GBMs have extensions, such as stochastic GBMs and extreme GBMs (XGBoost). If the basic model is well learned, various further hyperparameter designs are not difficult. For R implementation, the function `gbm()` is explained using the built-in non-time series data and the time series is then used.

Built-in data must be created. The `AmesHousing` package features the function `make_ames()`, which extracts the housing price data forAmes, Iowa. This is not a time series so the related functions of the `rsample` package are used to divide the data into training (`ames_train`) and testing (`ames_test`) subsets:

```
ames = AmesHousing::make_ames()
set.seed(123)
split = rsample::initial_split(ames, prop = 0.7, strata =
"Sale_Price")
ames_train = rsample::training(split)
ames_test = rsample::testing(split)
```

As shown in Fig. 3.20, this dataset has 81 variables and the target variable is `Sale_Price` (transaction price).

```
> colnames(ames)
 [1] "MS_SubClass"      "MS_Zoning"        "Lot_Frontage"        "Lot_Area"         "Street"
 [6] "Alley"            "Lot_Shape"        "Land_Contour"        "Utilities"        "Lot_Config"
[11] "Land_Slope"       "Neighborhood"     "Condition_1"         "Condition_2"      "Bldg_Type"
[16] "House_Style"      "Overall_Qual"     "Overall_Cond"        "Year_Built"       "Year_Remod_Add"
[21] "Roof_Style"       "Roof_Matl"        "Exterior_1st"        "Exterior_2nd"     "Mas_Vnr_Type"
[26] "Mas_Vnr_Area"     "Exter_Qual"       "Exter_Cond"          "Foundation"       "Bsmt_Qual"
[31] "Bsmt_Cond"        "Bsmt_Exposure"    "BsmtFin_Type_1"      "BsmtFin_SF_1"     "BsmtFin_Type_2"
[36] "BsmtFin_SF_2"     "Bsmt_Unf_SF"      "Total_Bsmt_SF"       "Heating"          "Heating_QC"
[41] "Central_Air"      "Electrical"       "First_Flr_SF"        "Second_Flr_SF"    "Low_Qual_Fin_SF"
[46] "Gr_Liv_Area"      "Bsmt_Full_Bath"   "Bsmt_Half_Bath"      "Full_Bath"        "Half_Bath"
[51] "Bedroom_AbvGr"    "Kitchen_AbvGr"    "Kitchen_Qual"        "TotRms_AbvGrd"    "Functional"
[56] "Fireplaces"       "Fireplace_Qu"     "Garage_Type"         "Garage_Finish"    "Garage_Cars"
[61] "Garage_Area"      "Garage_Qual"      "Garage_Cond"         "Paved_Drive"      "Wood_Deck_SF"
[66] "Open_Porch_SF"    "Enclosed_Porch"   "Three_season_porch"  "Screen_Porch"     "Pool_Area"
[71] "Pool_QC"          "Fence"            "Misc_Feature"        "Misc_Val"         "Mo_Sold"
[76] "Year_Sold"        "Sale_Type"        "Sale_Condition"      "Sale_Price"       "Longitude"
[81] "Latitude"
```

Fig. 3.20 Names of 81 variables for the data `Ames`

The `gbm` package is called directly, and the `gbm()` function is used to train the model, as follows:

```
library(gbm)
ames_gbm1 = gbm(
formula = Sale_Price ~.,
data = ames_train,
distribution = "gaussian", # SSE loss function
n.trees = 500,
shrinkage = 0.1, #learning rate
interaction.depth = 3,
n.minobsinnode = 10,
cv.folds = 10 # k of k-fold CV
)
```

The estimation result is stored in the object `ames_gbm1` and then `which.min` is used to determine the minimum value for the prediction error:

```
best = which.min(ames_gbm1$cv.error)
```

The RMSE is then calculated as:

```
sqrt(ames_gbm1$cv.error[best])
```

The tree that is used to predict `ames_test` is less than 1000, as shown in Fig. 3.21.

```
predict.err=predict(ames_gbm1, ames_test)-
                         ames_test[,"Sale_Price"]
```

The RMSE for the prediction that is generated by the GMB is then calculated and this result is compared with `sqrt(ames_gbm1$cv.error[best])`, which is much smaller, so unity is strength, as shown below:

```
> predict(ames_gbm1,ames_test)
Using 957 trees...

 [1] 124151.00 179543.69 195952.06 176730.80
[11] 161532.05 126956.48  88156.88 181481.25
[21] 188373.70 193451.14 255672.87 234921.85
[31] 172716.13 151536.96 164715.95 243032.28
[41] 178825.74 185483.73 135091.94  77253.11
[51] 130458.75 124229.57 142366.87 106884.98
[61] 201181.48 133241.17 128611.98 142623.82
[71] 124164.54 164079.02 159812.68 133419.21
[81] 143242.97  86829.59 111359.02 179158.45
[91] 126483.05 338648.65 140946.07 184803.35
```

Fig. 3.21 The predicted values that are generated by a GBM, using `ames` housing data

```
> sqrt(ames_gbm1$cv.error[best])
[1] 22536.9
> sqrt((sum(predict.err^2))/(nrow(ames_test)-1))
[1] 20849.53
```

The implementation of the RLab time series data now uses `train()`. The previous inflation data is used with a lag of 12 steps.

3.4.2 *RLab. Multistep Forecasts of GBM*

`iForecast` wrapper `tts.caret` is used, so the implementation is much easier using the code chunk:

Code Chunk

```
output1.gbm = tts.caret(y=y,
                        x=NULL,
                        arOrder=c(1:6),
                        xregOrder=c(0),
                        method="gbm",
                        train.end="2021-12-01",
                        type="none",
                        tuneLength =14,
                        resampling="boot",
                        preProcess = "center")
```

Specific dummies are included, so two models are forecast and Fig. 2.22 plots the results.

Finally, data analysis is used to calculate the importance of each variable using the command:

```
library(gbm)
varImp(output2.gbm$output)
```

Since `varImp()` is a wrapper it calls a function that calculates relative importance, so `varImp()` is used and the package `gbm` is preloaded.

R Exercises

1. f GBMs require a large amount of data to generate trees, but a greater lag structure for the time series is not necessarily better. Compare the predictive performance for AR(6) with that for AR(12).
2. Referring to Fig. 3.22, the average of two models improves forecasting performance. Confirm this conjecture.

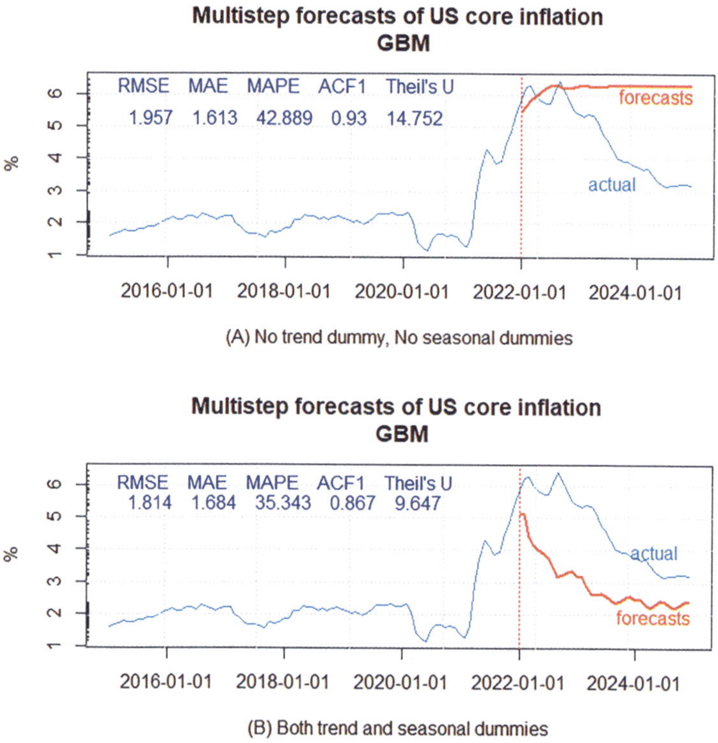

Fig. 3.22 Two time series plots that merge actual inflation with multistep forecasting values that are generated by *GBMs*. The solid thicker red line is the prediction. Performance measures for both models are similar but their directions are different: (**a**) shows the results for the model that uses no time dummies, which overpredicts the future, and (**b**) shows the results for the model that uses a trend and seasonal dummies, which underpredicts the future

3.5 Regularization Methods: LASSO, Ridge, and Elastic Net

3.5.1 Model Explanations

Regularization refers to the penalty that is assigned to extreme values of parameters. If the reader is already familiar with Regression Quantiles or Robust Statistics, the concept of a penalty should not be unfamiliar. The concept is described in terms of a linear regression OLS. A known linear regression is:

$$\mathbf{y} = \mathbf{X}\beta + \mathbf{e}$$

where $y = \{y_1, y_2,\ldots, y_n\}$ and $\beta = \{b_1, b_2,\ldots, b_p\}$. This regression is used with n observations and p feature variables, or exogenous, explanatory, independent variables.

This solution minimizes the sum of the squared residuals for the regression as:

$$\min .SSE = (y - X\beta)^{\mathrm{T}} (y - X\beta)$$

If $n > p$ and there is no (or slight) serious linear overlap, then by the Gauss–Markov theorem of classical statistics, least squares is the best estimate for the linear model. In many fields of machine learning, such as text mining or genetic analysis, it is common to violate $n > p$ for high-dimensional data and because of the large number of feature variables (big data), the model is difficult to interpret and there is unexpected multi-linear or -nonlinear correlation. If $n < p$, there is an almost infinite number of solutions for the least squares, so there is no unique and distinct solution.

In this case, a useful practice is to assume that there is a best smaller subset among the p feature variables and this subset integrates the strongest corresponding relationship. This strongest corresponding relationship is named by Hastie et al. (2017, p. 610) as the *bet on sparsity principle*. If the data is processed, feature selection techniques are used, such as the commonly used hard thresholding feature selection, which includes the forward selection and backward elimination that is common in traditional linear models. However, these methods are less efficient in terms of computation, as the name *hard thresholding* suggests, because they only judge according to the threshold in "addition" or "exclusion." Another method is called soft thresholding and gradually zeros the coefficients of irrelevant feature variables. Hard thresholding directly zeros when it fails the threshold test. In many practical analyses, the process of soft thresholding produces more accurate prediction models and results are easier to interpret.

Data for which $n < p$ is called wide data. For this type of data, there is no need to prove whether there is linear overlap. The alternative to LS for wide data is the regularization method, which is also known as the penalty or shrinkage method (see Hastie et al., 2017, pp. 61–73; Kuhn & Johnson, 2013, Chs. 6.4 and 12.5). The penalty method imposes restrictions to control the size of the estimated coefficients. Following the LS example, the objective function becomes

$$\min.\ SSE = \sum_{i=1}^{n} \left(y_i - \hat{y}_i\right)^2$$

St. $L \leq t$

L has two types:

$$L_1 = \text{LASSO Penalty} = \sum_{j=1}^{p} \left|\beta_j\right|$$

$$L_2 = \text{Ridge Penalty} = \sum_{j=1}^{p} \beta_j^2$$

A method called ENET (Elastic Net) combines LASSO[4] and Ridge. The formal representation of the Lagrange function is:

$$\text{LASSO Penalty} = \underset{\beta}{\arg\min}\left(\sum_{i=1}^{n}\left(y_i - \hat{y}_i\right)^2 + \lambda_1 \sum_{j=1}^{p}\left|\beta_j\right|\right)$$

$$\text{Ridge Penalty} = \underset{\beta}{\arg\min}\left(\sum_{i=1}^{n}\left(y_i - \hat{y}_i\right)^2 + \lambda_2 \sum_{j=1}^{p}\beta_j^2\right)$$

$$\text{ENET} = \underset{\beta}{\arg\min}\left(\sum_{i=1}^{n}\left(y_i - \hat{y}_i\right)^2 + \lambda_1 \sum_{j=1}^{p}\left|\beta_j\right| + \lambda_2 \sum_{j=1}^{p}\beta_j^2\right)$$

where λ is the standard symbol for Lagrange multipliers. Currently, in machine learning literature, similar regularization penalty methods also include least angle regression (LAR). Interested readers can refer to works such as that by Hastie et al. (2017).

In R, there are several packages that implement these three regularization methods. `MASS::lm.ridge()` can implement ridge regression models and `elasticnet::enet()` can implement elastic net models.

In theory, Ridge and Lasso are special cases of elastic net. The related parameters in Ridge are controlled, e.g., lambda = 0 results in LASSO must declare lambda and a fraction. The previous AR(6) is used for the RLab, starting with LASSO.

3.5.2 *RLab*

To save space, the code chunk is not displayed in text, and interested readers should refer to the Chapter code file. Figures 3.23 and 3.24 show the results.

R Exercises

1. Complete the above built-in calculations and compare the predictive performance for the three regularization methods.

[4] LASSO = *Least Absolute Shrinkage and Selection Operator*. See Tibshirani (1996) for details.

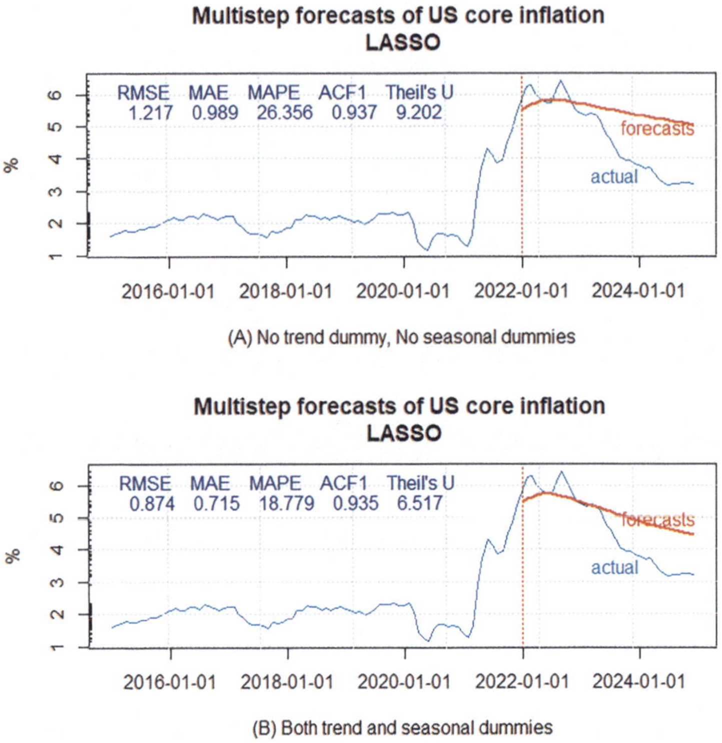

Fig. 3.23 Two time series plots that merge actual inflation with multistep forecasting values that are generated by *LASSO*. The solid thicker red line is the prediction: (**a**) shows the results for the model that uses no time dummies and (**b**) shows the results for the model that uses a trend and seasonal dummies. Performance measures for both models are similar

2. Use tuneLength to change the number of training values to 100 and then compare the predictive performance for the three models. Is a longer tuneLength better?

3.6 Automated Machine Learning Committee: autoML

3.6.1 Model Explanations

This section describes the automated machine learning solutions that are provided by the H_2O.ai platform. H_2O.ai is a professional cloud computing platform on which AI components and machine learning algorithms can be used. For example, H_2O.gbm is a GBM that is built in the H_2O environment. The components include cutting-edge machine learning algorithms, performance matrices, and auxiliary functions that make machine learning methods powerful and concise. On the H_2O.ai platform, automated algorithms are used to aggregate multiple machine learning

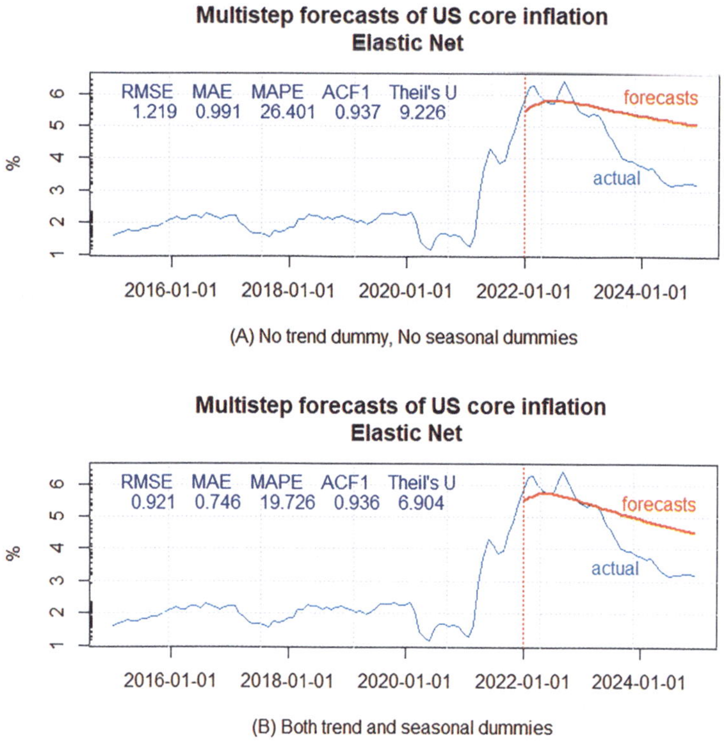

Fig. 3.24 Time series plot merging actual inflation with multistep forecasting values that are generated by *elastic net*. The solid thicker red line is the prediction: (**a**) shows the results for the model that use no time dummies and (**b**) shows the results for the model that uses a trend and seasonal dummies. Performance measures for both models are similar

models, including: *supervised deep learning (neural networks)*, *random forest*, *generalized linear models*, *gradient boosting machines*, *naive Bayes,* and *stacked ensembles*. After large-scale training, the best results for each model are combined to give the most suitable prediction (represented by autoML) using stacked reinforcement (stacked ensembles). In machine learning literature, this method is also known as the committee approach and is similar to the statistical forecasting average method at the end of Chap. 2. Therefore, the predicted values for autoML are not created using a single model; the comprehensive result for many models is used.

Currently, autoML is the most common architecture for machine learning. From the perspective of R, an explanation of how to install H_2O is necessary.

3.6.2 *Windows Installation*

H2O is not Windows-only and has up-to-date installation for Mac. It cannot be installed in Mac in 2023.

To use h2o.AutoML, its Java and R are installed via the link: https://h2o-release.s3.amazonaws.com/h2o/rel-3.46.0/6/index.html

This link has a string of version information: "/rel-3.46.0/6". This version was used to write this book. It is also directly accessed via the official homepage: https://h2o.ai/. Figure 3.25a shows the process to select **H2O AutoML** technical documentation.

Figure 3.25b shows two steps:

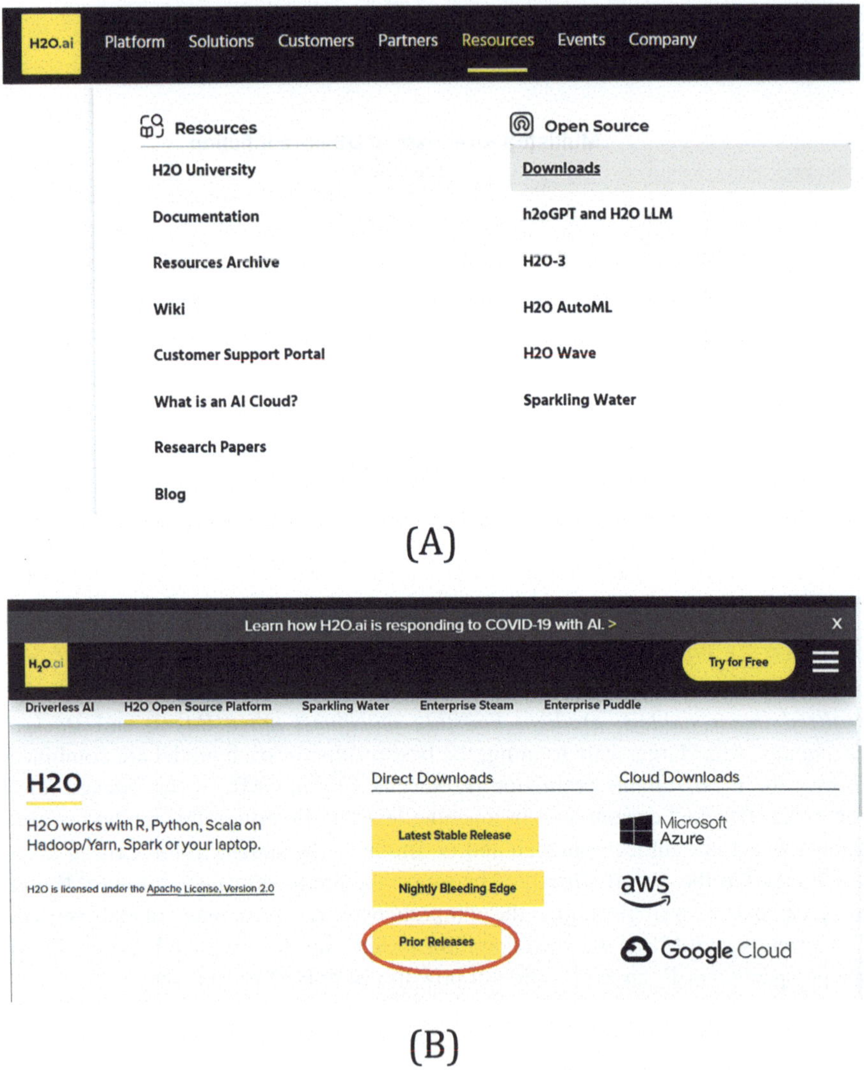

Fig. 3.25 Instructions on the platform: after successful installation, use the `tts.autoML()` function of `iForecast`. (**a**) Click **Downloads** and select button: **H2O Open Source Platform**, (**b**) enter the earlier versions, and (**c**) the web page of **H2O Open Source Platform**

Firstly, follow the instructions to install the Java of H2O. However, the version of H2O on CRAN must be consistent with the downloaded Java. The latest version is the Latest Stable Release in Fig. 3.25b, which is 3.46.0, and this is checked with h2o on CRAN. CRAN lags behind so prior releases must be entered to download the appropriate Java, or h20 can be installed via Amazon Hub. The lag is due to a time restriction between two updates' uploading to CRAN. Python does not have such a restriction. CRAN is less popular than PyPI with many data analyzers so the best way to install h2o is through h2o Amazon Hub: not CRAN.

Secondly, go to INSTALL IN R (INSTALL IN Python is also provided) and follow the instructions to complete the installation.

Finally, load H2O and start an H2O cluster:

```
library(h2o)
h2o.init()
```

Both commands initialize h2o and a successful connection message appears on the screen (Fig. 3.26). The autoML program can then be run using the `iForecast` wrapper `tts.autoML()` in this section. The h2o machine supports most machine learning, such as LASSO and GBM, which can be called individually.

3.6.3 RLab. Multistep Forecasts of autoML

In `iForecast`, there are two wrappers for time series forecasting using machine learning: `tts.caret` and `tts.autoML`. The code chunk below is easily to implement.

Code Chunk

```
out1.autoML=tts.autoML(y=y,
                       x=NULL,
                       arOrder=c(1:6),
                       xregOrder=c(0),
                       type= "none",
                       train.end="2021-12-01")
```

Since `tts.autoML` executes 20 algorithmic methods, it takes longer than `tts.carat`. The wrapper `tts.autoML` is easier to use, and the arguments are almost the same so the code details are omitted. The forecasting results are shown in Figs. 3.27 and 3.28. Figure 3.28 lists 22 algorithmic methods that are used in autoML.

To determine how many methods are used, the command below prints all commands on the screen:

```
print(out1.autoML$modelsUsed,
      n = nrow(out1.autoML$modelsUsed))
```

```
> h2o.init()

H2O is not running yet, starting it now...

Note:  In case of errors look at the following log files:
    C:\Users\badal\AppData\Local\Temp\Rtmp6peB2T\file70545dee5b3c/l
    C:\Users\badal\AppData\Local\Temp\Rtmp6peB2T\file705438d01f87/l

java version "1.8.0_431"
Java(TM) SE Runtime Environment (build 1.8.0_431-b10)
Java HotSpot(TM) 64-Bit Server VM (build 25.431-b10, mixed mode)

Starting H2O JVM and connecting:  Connection successful!

R is connected to the H2O cluster:
    H2O cluster uptime:          8 seconds 23 milliseconds
    H2O cluster timezone:        Asia/Taipei
    H2O data parsing timezone:   UTC
    H2O cluster version:         3.46.0.6
    H2O cluster version age:     3 months and 16 days
    H2O cluster name:            H2O_started_from_R_badal_ymh292
    H2O cluster total nodes:     1
    H2O cluster total memory:    7.01 GB
    H2O cluster total cores:     0
    H2O cluster allowed cores:   0
    H2O cluster healthy:         TRUE
    H2O Connection ip:           localhost
    H2O Connection port:         54321
    H2O Connection proxy:        NA
    H2O Internal Security:       FALSE
    R Version:                   R version 4.4.2 (2024-10-31 ucrt)
```

Fig. 3.26 Screenshot of a console output showing the initialization of an H2O cluster in R. If the installation of Java and R package `h2o` are successful, typing `h2o.ini()` shows the information on the screen. The process includes starting the H2O JVM and confirming a successful connection. Details include Java version 1.8.0_431, timezone Asia/Taipei, and cluster version 3.46.0.6

The next section describes how to perform a onestep ahead forecast using a moving window.

3.7 Additional Topics

3.7.1 Onestep Ahead Forecast

Onestep ahead prediction generates one out-of-sample forecasting value in terms of the last observations from a predictive regression. Using explanatory variables increases the forecasting performance but the forecast is limited to one period ahead. Many onestep forecasts use a moving window, as explained in Sect. 2.7.4 (Fig. 2.22) of Chap. 2. The key code for this is:

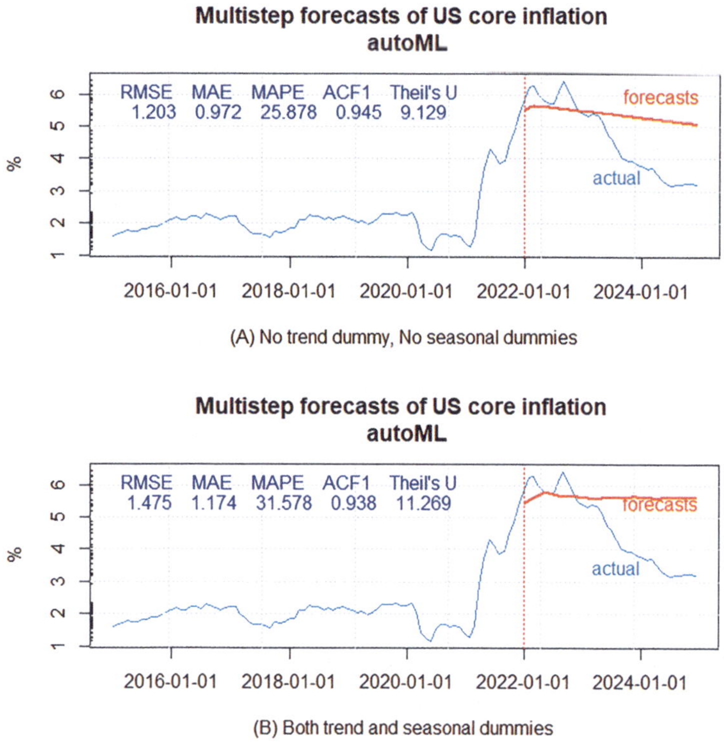

Fig. 3.27 Time series plot merging actual inflation with multistep predicted values that are generated by `tts.autoML`. The solid thicker red line is the prediction: (**a**) shows the results for the model that uses no time dummies and (**b**) shows the results for the model that uses a trend and seasonal dummies. Performance measures for both models are similar

Code Chunk. Data Manipulation

```
1.  tmp=read.csv("./data/CPIx.csv") #core CPI
2.  head(tmp)
3.  y0=timeSeries(tmp[,-1],as.Date(tmp[,1]))
4.  y=na.omit(diff(log(y0[,1]),12))*100
5.  INF=cbind(y,y0[-seq(12),-c(1)])
6.  colnames(INF)=c("inf", "spread", "unrate")
7.  dep=INF[,1]
8.  indep=INF[,-1]
9.  J=window(dep, start="2021-12-01", end=end(INF))
10.  J=rownames(J)
```

There are two points to explain in terms of data manipulation:

It is based on `timeSeries` object, which is slightly different to the `ts` object for base R: especially Line 9.

For `iForecast`, there is no need to lag the raw data. As shown in Line 6, the lag structure can be executed within the wrapper `tts.caret`. See Line 5 and Line 6 of the Code Chunk below.

	model_id	rmse	mse	mae	rmsle
1	StackedEnsemble_AllModels_1_AutoML_2_20250218_103746	0.231	0.054	0.163	0.057
2	StackedEnsemble_BestOfFamily_1_AutoML_2_20250218_103746	0.232	0.054	0.164	0.057
3	GLM_1_AutoML_2_20250218_103746	0.236	0.056	0.166	0.059
4	DeepLearning_grid_2_AutoML_2_20250218_103746_model_1	0.248	0.062	0.177	0.061
5	DeepLearning_grid_3_AutoML_2_20250218_103746_model_2	0.271	0.073	0.178	0.067
6	DeepLearning_grid_3_AutoML_2_20250218_103746_model_1	0.275	0.076	0.184	0.067
7	GBM_grid_1_AutoML_2_20250218_103746_model_1	0.300	0.090	0.200	0.069
8	GBM_5_AutoML_2_20250218_103746	0.302	0.091	0.203	0.067
9	GBM_4_AutoML_2_20250218_103746	0.306	0.093	0.202	0.068
10	XRT_1_AutoML_2_20250218_103746	0.308	0.095	0.203	0.068
11	GBM_3_AutoML_2_20250218_103746	0.312	0.097	0.205	0.068
12	DRF_1_AutoML_2_20250218_103746	0.314	0.099	0.206	0.068
13	GBM_grid_1_AutoML_2_20250218_103746_model_2	0.316	0.100	0.206	0.067
14	DeepLearning_grid_1_AutoML_2_20250218_103746_model_1	0.320	0.102	0.227	0.077
15	GBM_2_AutoML_2_20250218_103746	0.327	0.107	0.212	0.069
16	DeepLearning_grid_1_AutoML_2_20250218_103746_model_2	0.328	0.108	0.232	0.078
17	DeepLearning_1_AutoML_2_20250218_103746	0.333	0.111	0.249	0.085
18	GBM_grid_1_AutoML_2_20250218_103746_model_3	0.341	0.116	0.218	0.071
19	DeepLearning_grid_2_AutoML_2_20250218_103746_model_2	0.770	0.593	0.561	0.171
20	GBM_grid_1_AutoML_2_20250218_103746_model_5	0.875	0.765	0.480	0.124
21	GBM_grid_1_AutoML_2_20250218_103746_model_4	0.878	0.771	0.478	0.122
22	GBM_1_AutoML_2_20250218_103746	0.883	0.779	0.482	0.123

Fig. 3.28 List of 22 algorithmic methods that are used in autoML

Code Chunk. Loop for a moving-window estimation and a onestep forecast:

```
fcst1=NULL
for(i in 2:length(J)) {
output1 = tts.caret(y=dep,
                    x=indep,
                    arOrder=c(1:6),
                    xregOrder=c(1:2),
                    method="svm",
                    train.end=J[i-1],
                    type= "both",
                   tuneLength =10,
                   resampling="boot",
                   preProcess = "center")
 testData=output1$dataused[J[i],]
 fcst0=iForecast(Model=output1,
                  Type="static",
                newdata = testData)
 fcst1=c(fcst1,fcst0)
 print(i)
 }
 fcst1=timeSeries(fcst1, J[-1])
 Actual=window(dep, start=J[2],end=end(dep))
 forecast::accuracy(x=as.ts(Actual),as.ts(fcst1))
```

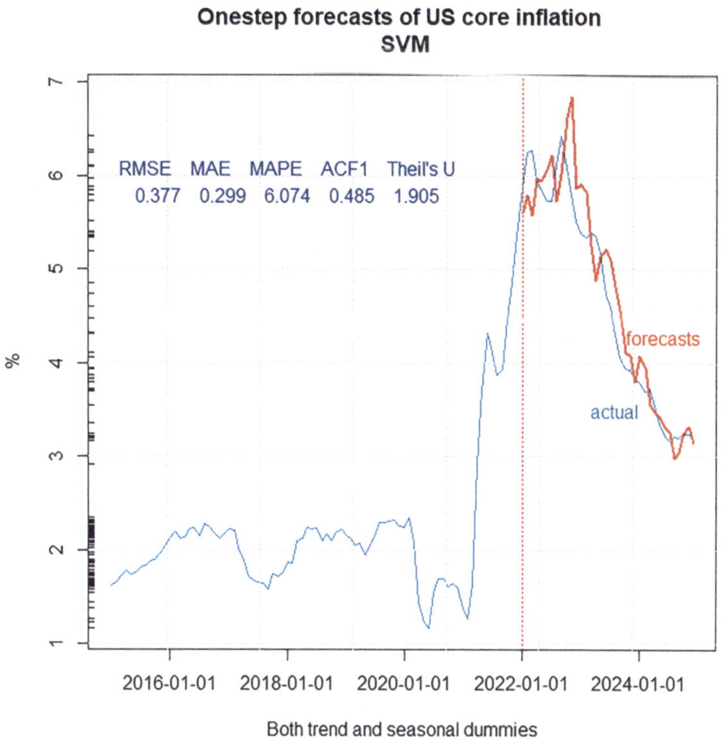

Fig. 3.29 Time series plot merging actual inflation with the values that are predicted using a SVM, which involves a moving-window onestep ahead forecast. The solid thicker red line is the prediction. The overall, prediction for this scenario is quite accurate

When the moving-window onestep loop is complete, the forecasting values are used to construct the `timeSeries` object (Line 20) and the forecasting accuracy measures are calculated (Line 22). The results are plotted in Fig. 3.29.

3.7.2 *Flexible Lag Structure of* `iForecast`

Data that is input to `iForecast` has two characteristics:

Similarly to a dynamic linear model, it requires raw data only. Lags can be specified easily inside.

Unlike ARIMA, non-sequential lags can be easily specified.

	y	ar1	ar4	ar7	spread_L3	unrate_L3	spread_L6	unrate_L6	spread_L9	unrate_L9
2024-07-01	3.163	3.224	3.727	3.836	-0.09	3.9	-0.11	3.7	0.01	3.9
2024-08-01	3.213	3.163	3.552	3.801	-0.08	4	-0.09	3.9	-0.06	3.7
2024-09-01	3.207	3.213	3.354	3.693	-0.09	4.1	-0.09	3.9	-0.09	3.8
2024-10-01	3.247	3.207	3.224	3.727	-0.13	4.2	-0.09	3.9	-0.11	3.7
2024-11-01	3.247	3.247	3.163	3.552	-0.28	4.2	-0.08	4	-0.09	3.9
2024-12-01	3.197	3.247	3.213	3.354	-0.41	4.1	-0.09	4.1	-0.09	3.9

Fig. 3.30 Data representing the non-sequential distributed lags for `iForecast`, which is viewed using `tail(output$dataused)`. To include *contemporaneous* feature variables (X_t variables), a 0 is added to `xregOrder`. For example, `xregOrder=c(0,3,6,9)`

Code Chunk

```
output = tts.caret(
        y=dep,
        x=indep,
        arOrder=c(1, 4, 7),
        xregOrder=c(3, 6, 9),
        method="svm",
        train.end="2021-12-01",
        type="none",
        tuneLength =14,
        resampling="boot",
        preProcess = "center")
tail(output$dataused)
```

As the code chunk above shows, the `arOrder` and `xregOrder` create the data structure that is shown in Fig. 3.30. To include *contemporaneous* feature variables (X_t variables), a 0 is added. For example, `xregOrder=c(0,3,6,9)`. However, if $p > 1$, $y_t = f(X_t, X_{t-p})$ does not generate a onestep ahead predictive value over time: its prediction is merely a model fitted value.

Chapter 4
Special Topics

4.1 VAR Forecasting Using Machine Learning

4.1.1 Model Explanations

VAR is an abbreviation for **V**ector **A**uto**r**egression model, which was proposed by the econometrician Sims (1980). VAR addresses the problem of specifying an equation $Y_t = \alpha + \beta X_t + u_t$ to study the bivariate causal relation. However, it confronts the challenges of possible facts of $X_t = \alpha + \beta Y_t + u_t$. That is, the dependent variable and the independent variable cannot be determined. An example is the relationship between consumption and income, which mutually affect each other. To some degree, VAR addresses this issue by conducting a Granger causality. VAR can also generate multistep forecasts with additional lagged regressors, similarly to AR.

Similarly to AR(p), VAR(p) represents a VAR system with p-order lags. If $Y_t = \alpha + \beta X_t + u_t$, vector $\mathbf{Z}_t = [X_t, Y_t]$, so VAR(1) is vectorized AR(1) as:

$$\mathbf{Z}_t = \alpha + \beta \mathbf{Z}_{t-1} + \mathbf{u}_t$$

The matrix representation of VAR(1) is:

$$\begin{bmatrix} Y_t \\ X_t \end{bmatrix} = \begin{bmatrix} \alpha_1 \\ \alpha_2 \end{bmatrix} + \begin{bmatrix} \beta_{11} & \beta_{12} \\ \beta_{21} & \beta_{22} \end{bmatrix} \begin{bmatrix} Y_{t-1} \\ X_{t-1} \end{bmatrix} + \begin{bmatrix} u_{1t} \\ u_{2t} \end{bmatrix}, \begin{pmatrix} u_{1t} \\ u_{2t} \end{pmatrix} \approx N(0, \Sigma)$$

The covariance matrix $\mathbf{\Sigma}$ is positive definite so for this VAR system, (X_t, Y_t) does not have concurrent dependency. Their correlation is reduced to the effects of lagged variables. Econometric explanation is not the focus of this book, which describes forecasting. If the estimation ends with the 100th period, forecasting consumption is used and income the 101st, 102nd, …, periods.

T.-w. Ho, *Time Series Forecasting using Machine Learning*,
https://doi.org/10.1007/978-3-031-97946-0_4

$$\begin{bmatrix} C_{100} \\ Y_{100} \end{bmatrix} = \begin{bmatrix} 8 \\ 9 \end{bmatrix}$$

$$\begin{bmatrix} C_{101} \\ Y_{101} \end{bmatrix} = \begin{bmatrix} 6.96 \\ 7.39 \end{bmatrix} + \begin{bmatrix} 0.008 & 0.461 \\ 0.232 & 0.297 \end{bmatrix} \begin{bmatrix} 8 \\ 9 \end{bmatrix} = \begin{bmatrix} 11.173 \\ 11.919 \end{bmatrix}$$

$$\begin{bmatrix} C_{102} \\ Y_{102} \end{bmatrix} = \begin{bmatrix} 6.96 \\ 7.39 \end{bmatrix} + \begin{bmatrix} 0.008 & 0.461 \\ 0.232 & 0.297 \end{bmatrix} \begin{bmatrix} 11.173 \\ 11.919 \end{bmatrix} = \begin{bmatrix} 12.544 \\ 13.522 \end{bmatrix}$$

$$\vdots$$

$$\vdots$$

As shown previously, including additional regressors can improve the univariate forecasting performance but it is limited to onestep ahead and cannot give prospects over more steps ahead. Modeling using VAR overcomes this problem. Conventionally, VAR is estimated using LS or MLE.

In line with previous methods, in the following RLab, a 3-variate VAR(1) for inflation is built and estimated and an out-of-sample forecast is generated using multistep.

$$\begin{bmatrix} \text{inflation}_t \\ \text{spread}_t \\ \text{unrate}_t \end{bmatrix} = \begin{bmatrix} a_1 \\ a_2 \\ a_3 \end{bmatrix} + \begin{bmatrix} b_{11} & b_{11} & b_{11} \\ b_{11} & b_{11} & b_{11} \\ b_{11} & b_{11} & b_{11} \end{bmatrix} \begin{bmatrix} \text{inflation}_{t-1} \\ \text{spread}_{t-1} \\ \text{unrate}_{t-1} \end{bmatrix}$$

After statistical forecasts, VAR is estimated and forecast VAR using machine learning.

4.1.2 RLab. Forecasting VAR Using Machine Learning

The dataset INF that is compiled in Sect. 3.7.1 is used.

Code Chunk. Baseline VAR using LS

```
library(vars)
insample=window(INF,
           start=start(INF),
           end="2021-12-01")
outsample=window(INF,
           start="2022-01-01",
           end=end(INF))
out.var=VAR(insample, lag.max=12, type="both")
h= nrow(outsample)
fcst_var=predict(out.var,
                  n.ahead=h)[[1]]$inf[,1]
fcst_var=ts(fcst_inf,start=c(2022,1),freq=12)
accuracy(x=as.ts(outsample[,1]),fcst_var)
```

Line 4 estimates the econometric VAR and `predict(out.var)` retrieves the out-of-sample forecasts for three variables. Line 6 extracts forecasts.

The following code chunk applies machine learning to a VAR system of equations. The example is a SVM.

Code Chunk. VAR by Machine Learning

```
1.  OUT1=tts.var(data=INF,
                 p=18,
                 method="svm",
                 train.end="2021-12-01",
                 type="trend")
2.  OUT2=iForecast.var(OUT1, n.ahead=nrow(outsample))
3.  rownames(OUT2)=rownames(outsample)
4.  OUT2=as.timeSeries(OUT2)
5.  accuracy(x=as.ts(outsample[,1]),as.ts(OUT2[,1]))
```

Line 1. The wrapper `tts.var` is straightforward and calls `tts.caret`. It is time-consuming so it does not support `tts.autoML`.

Line 2. The forecasting function for the `tts.var` object, which is built as a multistep forecast, so only the argument `n.ahead` is declared, similarly to the `ahead` at Chap. 3.

The remaining three Lines are easily understood. Figure 4.1 puts both VARs together to make comparison easier. VAR-SVM performs no worse than `auto.arima` in Fig. 2.10, which uses a first-order difference and more ARMA terms and does not difference inflation. A simple ARMA does not give satisfactory performance. In general, univariate machine learning performs better without pre-differencing. Machine learning VAR gives a better predictive result than LS-VAR.

The optimal lag term for LS-VAR is 10, and the value for a SVM-VAR is 18. Machine learning involves the time-consuming trials of different lag terms but even if parallel computation is used, searching the best result is less efficient.

R Exercises

1. Conduct a Diebold–Mariano test to determine the predictive accuracy of forecasting values for Figs. 2.10 and 4.1b.
2. Try different method for `tts.var`, such as random forest and GBM. Discuss the outcomes.

4.2 Discrete-Valued Dependent Variable

A time series is an arranged sequence of observations in time order, which is assumed to be part of the realization of an underlying stochastic process. Since the pioneering textbook by Box and Jenkins (1970), time series analysis has become a well-established discipline. Although most time series are assumed to be real-valued, a discrete-valued dependent variable, such as count time series, binary

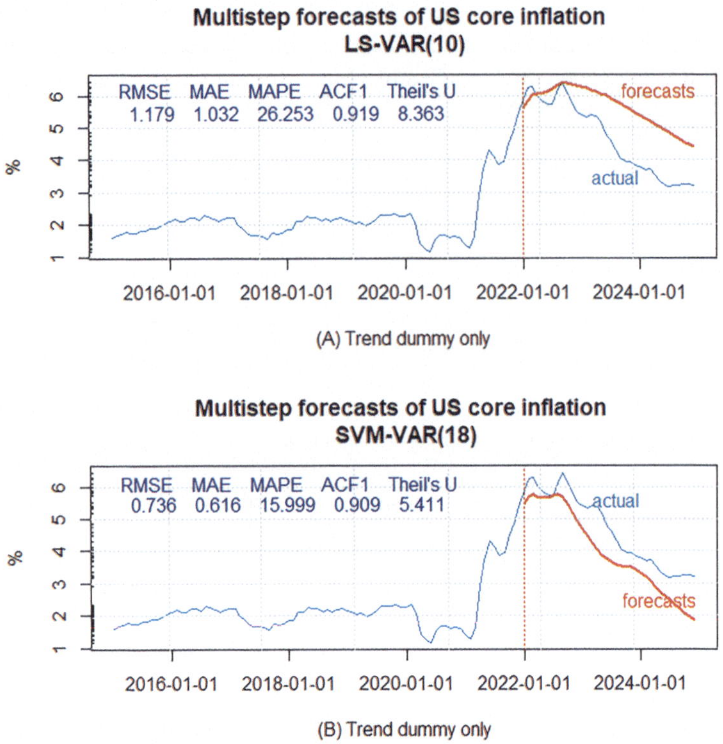

Fig. 4.1 Multistep forecasts compared: (**a**) Results for VAR(10) that are estimated using a conventional LS method. The forecasting values over-predict inflation and the RMSE is 1.18. (**b**) shows the results using a SVM to estimate VAR(18) and the resulting multistep forecasts are quite good, with a RMSE = 0.736. Compared with Fig. 2.10, it is similar to ARIMA(RMSE = 0.758). Machine learning does not return the same result but forecasting VAR using machine learning is a viable alternative

series, or a qualitative time series with a nominal or ordinal range are possible. In these cases, the common models and methods for real-valued time series cannot be used.

As Figs. 3.5 and 3.6 show, if the target variable is categorical, it is discrete-valued and it is statistically known as a generalized linear model. Its econometric counterpart is a LIMDEP (limited dependent variable). The most common case is the binary choice that is described by the Bernoulli trial {failure, success}. A binary dependent variable is generalized to multinorm in an extensive sense of classification. In terms of data mining, the {failure, success} string is also known as labels, which are recoded to {0, 1} for computation.

Chapter 3 briefly describes the input data table for classification, which is cross-section. In terms of statistical time series, interested readers are referred to Davis et al. (2016) for theoretical details, and Weiss (2018) offers a very good introduction to integer-value time series. From the data-driven classification perspective, Maharaj et al. (2019) is a good reference for machine learning.

	y	IPI_TWN	LD_OECD	LD_G7	LD_NAFTA
1989-01-01	2	-0.04017	-0.0002	-0.0003	-0.00099
1989-02-01	2	-0.24725	-0.0004	-0.00069	-0.00199
1989-03-01	2	0.28512	-0.00069	-0.00089	-0.001
1989-04-01	2	-0.03574	-0.00079	-0.00109	-0.002
1989-05-01	2	0.02725	-0.00079	-0.00099	-0.001
1989-06-01	1	-0.01353	-0.0007	-0.0008	-0.002
	...	...	...	...	...
2019-07-01	2	0.05876	-0.0004	-0.00091	-0.00091
2019-08-01	2	0.0174	0	-0.0002	0.0002
2019-09-01	2	-0.04051	0.0002	0	-0.0005
2019-10-01	2	0.02733	0.0005	0.0004	0.0005
2019-11-01	2	0.00333	0.00141	0.00141	0.00091
2019-12-01	2	0.00541	0.00091	0.00091	0.00181

Fig. 4.2 Data table for `data(bc)`. For a classification model, the target variable must be integer-valued, have no character and no factor. `iForecast` automatically transforms these into a factor level if limited discrete values are detected

In the package, `iForecast`, `tts.caret()` and `tts.autoML()` automatically check the attributes of a dependent variable. In a tree model, if the target variable is discrete-valued, then it will call a decision tree algorithm for classification analysis; otherwise, a regression tree is used.

The example uses a built-in dataset `bc`, which contains monthly business cycle data. y is a binary indicator(1 = recession, 2 = boom) for Taiwan's business cycle phases, IPI_TWN is the industrial production index for Taiwan, LD_OECD, LD_G7, and LD_NAFTA are leading indicators of OECD, G7, and NAFTA regions. All four continuous variables are monthly rates of change. Using `data(bc)` the data is loaded, and its contents are shown in Fig. 4.2. The target variable must be integer-valued, have no character and no factor. The package `iForecast` automatically transforms these into a factor level if limited discrete values are detected.

Code Chunk. Estimation of Classification Model

```
output1.svm = tts.caret(y=bc[,1],
                        x=bc[,-1],
                        arOrder=c(1:3),
                        xregOrder=c(1:6),
                        method="svm",
                        train.end="2017-12-01",
                        type="both,
                        tuneLength =14,
                        resampling="boot",
                        preProcess = "center")
```

Code Chunk. Prediction and confusion matrix

```
1. 1.  testData=window(output1.svm$dataused,
                         start="2018-01-01",
                         end=end(bc))
2.  fcst1.svm=iForecast(Model=output1.svm,
                         Type="static",
                         newdata = testData)
3.  fcst1.svm=as.timeSeries(fcst1.svm,
                         time(testData))
4.  tbl= table(Actual=testData[,1],
                         Prediction=fcst1.svm)
5.  caret::confusionMatrix(tbl)
```

For a discrete-valued dependent variable, accuracy measures, such as RMSE/MAPE,are not suitable. Line 4 uses the function `confusionMatrix` in the package `caret` to calculate the accuracy to determine the performance for a discrete-valued target. Figure 4.3 shows the results of Line 5.

For simplicity, this example simplifies the procedure for a onestep forecast by ignoring the moving window.

```
Confusion Matrix and Statistics

      Prediction
Actual  1  2
     1 12  1
     2  1 10

               Accuracy : 0.9167
                 95% CI : (0.73, 0.9897)
    No Information Rate : 0.5417
    P-Value [Acc > NIR] : 0.00008909

                  Kappa : 0.8322

 Mcnemar's Test P-Value : 1

            Sensitivity : 0.9231
            Specificity : 0.9091
         Pos Pred Value : 0.9231
         Neg Pred Value : 0.9091
             Prevalence : 0.5417
         Detection Rate : 0.5000
   Detection Prevalence : 0.5417
      Balanced Accuracy : 0.9161

       'Positive' Class : 1
```

Fig. 4.3 Confusion matrix results for Line 5. The Pos Pred Value 0.9231 is also known as precision. Precision equals sensitivity. On the top, the diagonal elements of the confusion matrix show that a SVM predicts both states (1 = recession, 2 = boom) equally well

4.3 Actual Future

In practice, the real future is of most interest. Training informs model selection and when the best model is selected,[1] the real future is forecast.

Both `tts.var()` and `tts.caret()` generate multistep forecasts for the actual future. The only key argument is the number of steps ahead and declaring the end of dataset as `train.end` without declaring the argument for `newdata` with respect to the corresponding forecasting function.

This is computationally burdensome for `tts.var`, especially if the data input has many variables and the time series is long, such as a 30 year weekly foreign exchange rate for ten major currencies. Therefore, caution is required while using `tts.var`. The current version of `iForecast` does not support parallel computation.

It is also possible to forecast the real future by specifying the predictive regression as:

$$y_t = a + bX_{t-5}$$

so $\hat{y}_{t+1},\dots\hat{y}_{t+5}$ is forecast by inserting $X_{t-4}, \dots X_t$. If X is a matrix of three variables, then $X_{t-4}, \dots X_t$ are realized observations so it is only necessary to create a 5×3 matrix as `newdata` within the function `iForecast()`, and declare a `Type="static"` to filter the prediction. The lag structure must be in accordance with the data that is used for estimation. Given its empirically poor forecasting performance, this is left as an exercise for the reader.

[1] For example, given a method, like random forest, the lag structure and time dummies are optimally determined.

Chapter 5
Predictive Case Studies: Training by Rolling

In previous chapters, the prediction is evaluated by separating two segments over time. The left segment is for estimation, and the right segment, which is the future, is for prediction. To determine the robustness of a model, its performance is measured by rolling over time.

As shown in Fig. 5.1a, the multistep forecasts uses: $h = 6$, each estimation window generates six out-of-sample forecasting values, which is called validation, and the estimation is update to generate six forecasting values. The process does not stop until the validation reaches the end of time. If it moves five times and the stacked time series of 30(T) forecasting values, which is denoted by $\hat{y}_{1,T}$, is the multistep forecast for Model 1, the accuracy measures for $\hat{y}_{1,T}$ are calculated. These are the RMSE and MAPE. Figure 5.1b shows the onestep forecast for $h = 1$.

This process is repeated to obtain k model-specific forecasting time series $\hat{y}_{1,T}, \hat{y}_{2,T}, \ldots, \hat{y}_{k,T}$. k forecasting series are created using k forecasting models, so the accuracy measure can be calculated and a Diebold–Mariano test can be performed to compare predictive accuracy.

5.1 Predictive Modeling of Unemployment Rate

If the date is 2024/12/31 and the goal is to forecast a 6-month ahead unemployment rate for 2025/1–2025/6, the best model is selected by training several models over rolling windows and all 6-step ahead forecasts are stacked. The accuracy measures and the Diebold–Mariano statistic for these six forecasts are then compared.

T.-w. Ho, *Time Series Forecasting using Machine Learning*,
https://doi.org/10.1007/978-3-031-97946-0_5

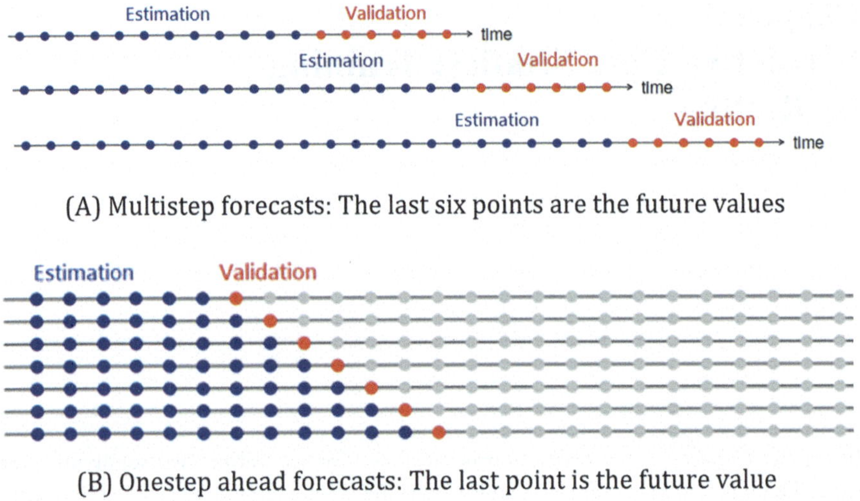

Fig. 5.1 For an estimation method and model, the estimation window rolls over time and stacks forecasting values for all moving validation segments: (**a**) Multistep forecasts: The segment consisting of the last six points is the future value and (**b**) Onestep ahead forecasts: The segment uses the last point as the future value. To determine whether it is a good forecasting constituent, the case studies are used to calculate the accuracy measures for the stacked forecasting values, and a pairwise Diebold–Mariano test is performed

5.1.1 Data and Model

The dataset consists of monthly observations for the period 1968/1–2024/12 for unemployment rate and three covariates: IC (Initial Claims for unemployment insurance), inflation rate (seasonal growth rate for CPI), and industrial growth (seasonal growth rate for industrial production index). IC is a common indicator for unemployment. Montgomery et al. (1998) considered it in the multivariate case. Inflation and industrial growth were considered in Edlund and Karlsson (1993) as economic control variables. All are obtained from the Fred[1] database, and the starting month is trimmed to match initial claims.

Figure 5.2 plots the time series. The COVID-19 pandemic resulted in a significant increase to 15% after 2020/3, and this figure then quickly reduced to around 8%. Figure 5.3 shows the dynamic relationship between initial claims (in millions) and the unemployment rate. The initial claims provide an indication of whether unemployment is improving or worsening and typically lead the unemployment rate by one or two quarters.

To implement a rolling window, `iForecast` has a built-in function `rollingWindows()`,which generates two sets of timestamps to define the start

[1] https://fred.stlouisfed.org/

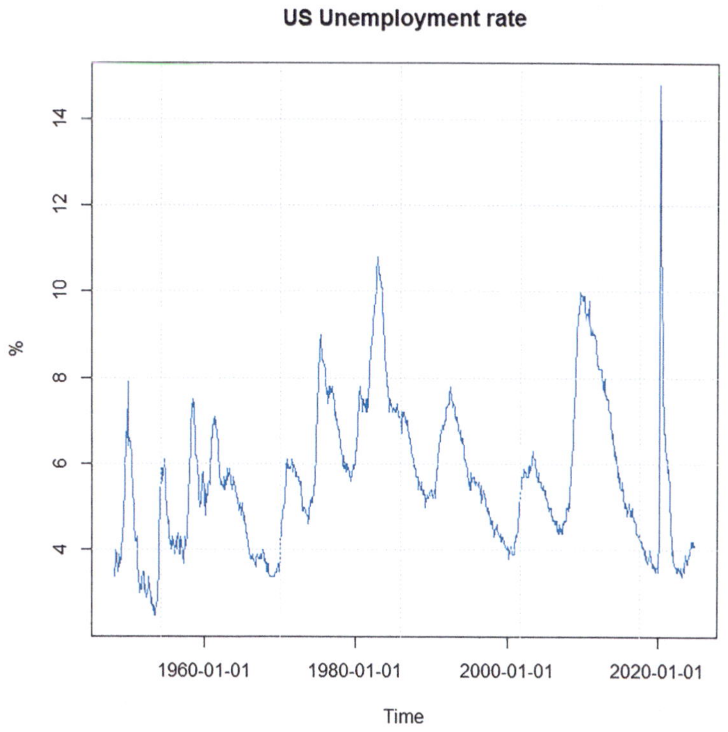

Fig. 5.2 The time series for US unemployment rate. The COVID-19 pandemic results in a significant increase to 15% after 2020 March, which quickly reduces to around 8%

and end of time segments, as shown in Fig. 5.1. The Code Chunk below shows the process for `rollingWindows`.

Code Chunk. The use of `rollingWindows`

```
library(timeSeries)
load("data/unrate.RData")
y=na.omit(unrate[,1])
tf= rollingWindows(y, estimation="720m", by="6m")
FROM=as.character(tf$from)
TO=as.character(tf$to)
```

The top three Lines define the US unemployment rate time series as y and last three Lines use `rollingWindow` to generate a timeframe(tf) and to extract two timestamp objects to define `FROM` and TO. Figure 5.4 shows the results.

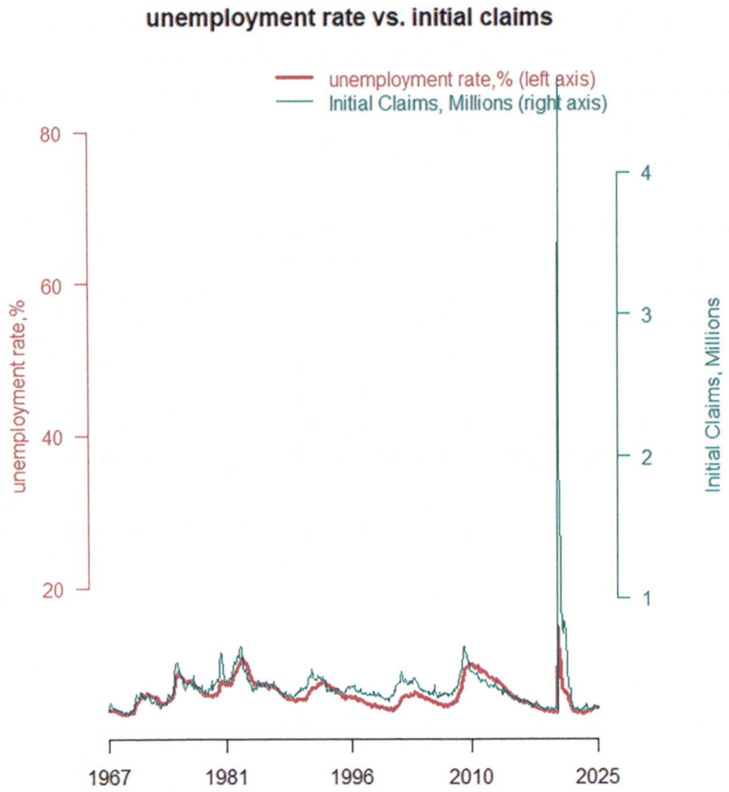

Fig. 5.3 The dynamic relationship between initial claims (in millions) and the unemployment rate. The initial claims provide an indication of whether unemployment is improving or worsening and typically lead the unemployment rate by 3 or 6 months

5.1.2 Univariate Multistep Forecasts

ARIMA Baseline

In the Code Chunk below:

Line 1 is the h-step ahead for multistep forecasts.

Line 2-Line12 is the loop for rolling windows and the indented block at Line 4-Line 11 are the looping contents.

Line 4 defines the EST as the unemployment rate for the estimation window, or in-sample data, and fixes the origin using `start=FROM[1]`. Line 5 defines VLD as that of validation window, or out-of-sample data. At the end of this line, the symbol [−1,] removes the first observation, which overlaps with the last observation of EST.

```
> timeframe
$from
 [1] "1948-01-01" "1948-07-01" "1949-01-01" "1949-07-01" "1950-01-01"
 [6] "1950-07-01" "1951-01-01" "1951-07-01" "1952-01-01" "1952-07-01"
[11] "1953-01-01" "1953-07-01" "1954-01-01" "1954-07-01" "1955-01-01"
[16] "1955-07-01" "1956-01-01" "1956-07-01" "1957-01-01" "1957-07-01"
[21] "1958-01-01" "1958-07-01" "1959-01-01" "1959-07-01" "1960-01-01"
[26] "1960-07-01" "1961-01-01" "1961-07-01" "1962-01-01" "1962-07-01"
[31] "1963-01-01" "1963-07-01" "1964-01-01" "1964-07-01" "1965-01-01"

$to
 [1] "2007-12-01" "2008-06-01" "2008-12-01" "2009-06-01" "2009-12-01"
 [6] "2010-06-01" "2010-12-01" "2011-06-01" "2011-12-01" "2012-06-01"
[11] "2012-12-01" "2013-06-01" "2013-12-01" "2014-06-01" "2014-12-01"
[16] "2015-06-01" "2015-12-01" "2016-06-01" "2016-12-01" "2017-06-01"
[21] "2017-12-01" "2018-06-01" "2018-12-01" "2019-06-01" "2019-12-01"
[26] "2020-06-01" "2020-12-01" "2021-06-01" "2021-12-01" "2022-06-01"
[31] "2022-12-01" "2023-06-01" "2023-12-01" "2024-06-01" "2024-12-01"
```

Fig. 5.4 The timeframe that is created by `rollingWindows`. Using two sub-objects for illustration, `from[1]` ~ `to[1]` defines the first time interval for the estimation window (in sample) and `to[1]` ~ `to[2]` defines the first time interval for the validation window (out-of-sample)

Line 6–Line11 show the standard procedures, as defined in Chap. 2. There is a `[seq(nrow(VLD))]` at the end of Line 7, which is used to match the number of observations for the last loop with the number of actual values. Although `hstep` and the length of `TO` are divisible for this example, fitting general circumstances requires this information.

Line 13 extracts the actual unemployment rate, Line 14 combines the forecasts and actual values, and these are saved as the `path` indicates.

Code Chunk. ARIMA baseline

```
1. hstep=6
2. FCST0=NULL
3. for (i in 1:(length(TO)-1)) {
    4. EST=window(y, start=FROM[1], end=TO[i])
    5. VLD=window(y, start=TO[i], end=TO[i+1])[-1,]
    6. out=auto.arima(EST,ic="aic")
    7. fcst=forecast(out, h=hstep)$mean[seq(nrow(VLD))]
    8. fcst= as.matrix(fcst)
    9. rownames(fcst)=rownames(outsample)
    10. fcst=as.timeSeries(fcst)
    11. FCST0 <- rbind(FCST0,fcst)
12. }
13. Actual=window(y, start=start(FCST0),end=end(FCST0))
14. FCST=cbind(Actual=Actual, FCST=FCST0)
15. path="./output/data/univariate/arima.RData"
16. save(y,FCST, file=path)
```

In Sect. 5.1, machine learning is used to determine the accuracy as:

```
     RMSE       MAE      MAPE      ACF1 Theil's U
    1.355     0.481     6.853     0.700     1.273
```

Fig. 5.5 The forecasting performance for ARIMA using a rolling window: the h-step is six and validation window moves from 2008/1 to 2024/12 to give a total of 204 forecasting values

```
accuracy(x=as.ts(FCST[,"Actual"]),as.ts(FCST[,"FCST"]))
```

which presented as Fig. 5.5.

Machine Learning

Chapter 3 shows that a training by rolling framework is basically a loop, which is illustrated by the following Code Chunk in Line 1 and Line 2.

Code Chunk. Training by Rolling Machine Learning

```
hstep=6
FCST0=NULL
  1.  for (i in 1:(length(TO)-1)) {
output = tts.caret(y=y,
                   x=NULL,
                   arOrder=c(1:12),
                   xregOrder=c(0),
                   method="svm",
                   train.end=TO[i],
                   type= "both,
                   tuneLength =14,
                   resampling= "boot",
                   preProcess = "center")
dataUsed = output$dataused
VLD=window(dataUsed, start=TO[i], end=TO[i+1])[-1,]
fcst=iForecast(Model=output,
               Type="dynamic",
               n.ahead = nrow(VLD))
fcst=as.timeSeries(fcst, as.character(time(VLD)))
FCST0=rbind(FCST0, fcst)
  2.  }
Actual=window(y, start=start(FCST0),end=end(FCST0))
FCST=cbind(Actual=Actual, FCST=FCST0)
path="./output/data/univariate/svm.RData"
save(y,FCST, file=path)
```

If machine learning is used, a more-than-one method is necessary. For example, three methods are applied: `"enet"`, `"gbm"` and `"rf"`. A normal solution is to add a method loop outside, as explained in the following:

Code Chunk. Adding Loop over Method

```
  1.  METHODS=c("enet","gbm","rf")
hStep=6
  2.  for (m in METHODS) {
          FCST0=NULL
          for (i in 1:(length(TO)-1)) {
          output = tts.caret(y=y,  x=NULL, arOrder=c(1:12),
                                      xregOrder=c(0),
  3.                                  method=m,
                                      train.end=TO[i],
                                      type= "both",
                                      resampling= "boot")
          dataUsed = output$dataused
          VLD=window(dataUsed, start=TO[i], end=TO[i+1])[-1,]
          fcst=iForecast(Model=output,
                                      Type="dynamic",
                                      n.ahead = nrow(VLD))
          fcst=as.timeSeries(fcst,as.character(time(VLD)))
          FCST0=rbind(FCST0,fcst)
          }
Actual=window(y,start=start(FCST0),end=end(FCST0))
FCST=cbind(Actual=Actual,FCST=FCST0)
  4.  PATH=paste0("./output/data/univariate/",m,".RData")
save(y, FCST, file=PATH)
  5.  }
```

Line 1 defines an object as a set of three methods.
Line 2 initializes the loop index *m*.
Line 3 links *m* to an argument method.
Line 4 constructs a *PATH* to save the results.
Line 5 is the end of this outer loop.

This loop depicts a sequential process, so `"rf"` is not implemented until the previous method (`"gbm"`) is finished. This process is inefficient and time-consuming if one of the methods requires a longer time, for example, random forest, xgbTree, and xgboost. These methods are computationally intensive. Parallel computation assigns each method to the computer memory as an independent job and is much more efficient.

This section only shows the naive parallel and does not address complex parallel computation. This involves two steps:

The added method loop (Line 1, Line 2, and Line 5) is removed, and Line 3 is retained. The re-written code is the script file `Ch5_uni4pll.R`.

The `m` in Line 3 is referenced to parallel, as detailed in the script file: `Ch5_pll.R`.

A simple illustration is shown in the following Code Chunk.

Code Chunk. Naïve Parallel: Ch5_pll.R

```
library(rstudioapi)
M=c("svm","enet","gbm","rf")
m=M[1];jobRunScript("Ch5_code4pll.R", importEnv = TRUE)
m=M[2];jobRunScript("Ch5_code4pll.R", importEnv = TRUE)
m=M[3];jobRunScript("Ch5_code4pll.R", importEnv = TRUE)
m=M[4];jobRunScript("Ch5_code4pll.R", importEnv = TRUE)
```

Figure 5.6 shows the screen displaying the previous Code Chunk.

At the end of the code, the results are saved and analyzed in terms of performance. As shown in Fig. 5.7, the RMSE values show that SVM is the only machine learning method that is better than ARIMA. However, in terms of the MAE, they are almost the same.

A Diebold–Mariano test is left as an exercise to reconcile these measures. Visualization is also important to check their tracking performance.

Figures 5.8 and 5.9 plot the forecasting values for four methods against actual values. The plot shows that from 2008 to 2020, all methods' forecasting performance is similar and the COVID-19 incident tests robustness. The SVM and ARIMA are stable but the other three machine methods become volatile after 2022.

Figure 5.7 shows that if the Diebold–Mariano statistic is calculated to justify the predictive accuracy, the testing scenario is a one-sided hypothesis for which the null

Fig. 5.6 The use of background jobs for naïve parallel computation. Four progress bars are displayed, each labeled "Chapter5_code6pl1.R," indicating ongoing processes. All progress bars are nearly complete, suggesting tasks are close to finishing

	RMSE	MAE	MAPE	ACF1	Theil's U
arima	1.355	0.481	6.853	0.700	1.273
enet	1.386	0.625	11.084	0.719	1.445
gbm	1.477	0.663	11.460	0.726	1.611
rf	1.812	0.861	16.977	0.742	2.326
svm	1.240	0.504	7.476	0.741	1.192

Fig. 5.7 Accuracy measures for all results. In terms of the RMSE and Theil's U, the SVM is the only machine learning method that outperforms ARIMA

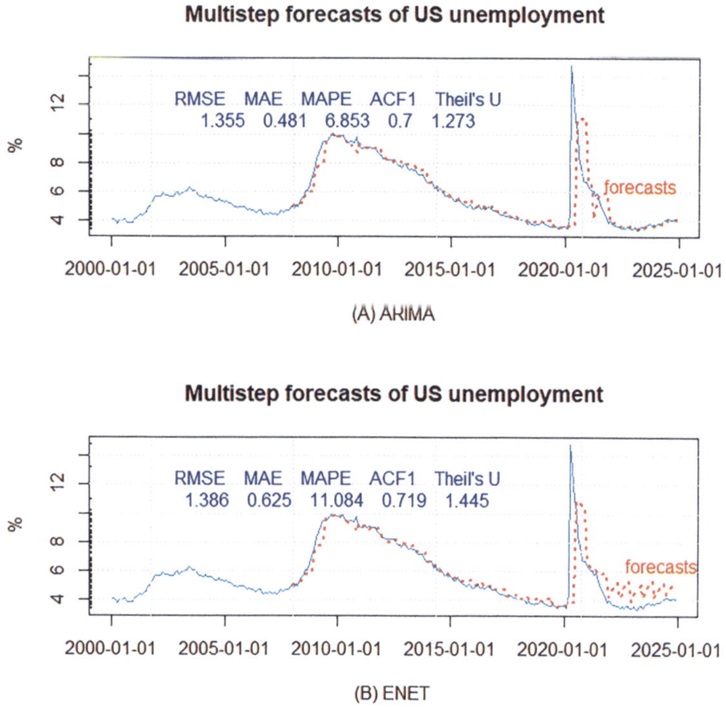

Fig. 5.8 Multistep forecasts using ARIMA and ENET and rolling windows. This graph plots the forecasting results for two methods: (**a**) ARIMA and (**b**) ENET. From 2008/1 to 2019/12, both methods perform equally and the COVID-19 incident tests robustness. ENET becomes volatile after 2022, and ARIMA remains stable

H_0 is equal accuracy and the alternative H_A has a smaller loss function. This test determines whether the null hypothesis can be rejected using the method that is specified by the alternative. Using Table 5.1 as an illustrative example, the format for H_0 is an ordered expression of errors for two models such that H_0: B = A. The alternative H_A implies that Model A outperforms Model B. If the loss function is the RMSE, Table 5.1 shows that over the full forecasting period, 2008/1–2024/12, the SVM outperforms the other four methods. Repeating the process without the SVM, ARIMA outperforms the other three methods.

Table 5.2 lists the Diebold–Mariano test results for two subperiods: 2008/1–2019/12 and 2020/1–2024/12. The loss function is the RMSE. The subperiods, respectively, correspond to the pre- and post-COVID periods. The pre-COVID test results show that ARIMA is the best forecasting constituent and for the post-COVID period, the SVM outperforms the other four methods and the next best model is ARIMA.

In summary, random forest is the worst forecaster.

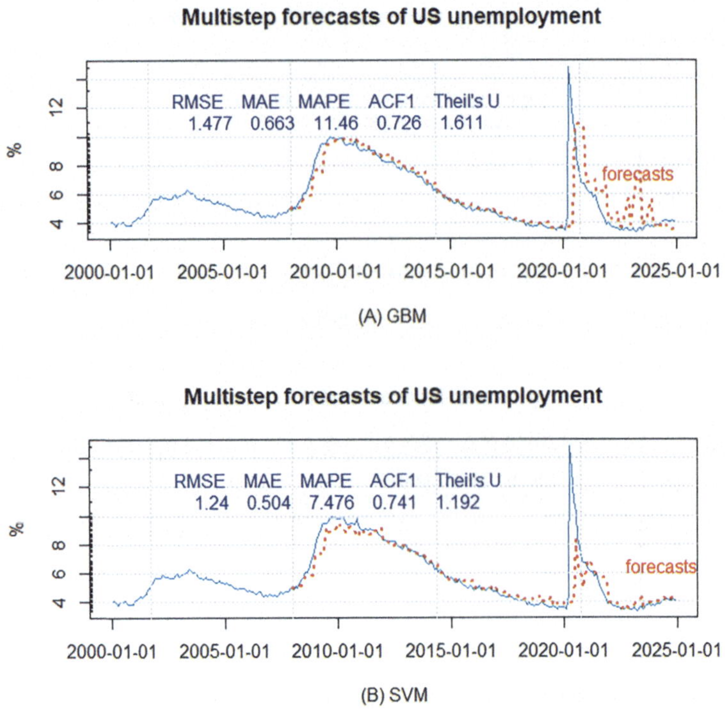

Fig. 5.9 Multistep forecasts for a GBM and a SVM using rolling windows. This graph plots the forecasting results for two methods: (**a**) GBM and (**b**) SVM. From 2008/1 to 2019/12, both methods perform equally and the COVID-19 incident tests robustness. The GBM becomes volatile after 2022, and the SVM remains stable

Table 5.1 Diebold–Mariano test for the full forecasting period, 2008/1–2024/12, for which the loss function is the RMSE. The SVM outperforms the other four methods, and the next best model is ARIMA

	2008/1–2024/12	
H_0	Statistic	*p*-value
ARIMA = **SVM**	1.8853	0.03
ENET = **SVM**	3.3118	0.0005
GBM = **SVM**	3.6489	0.0002
RF = **SVM**	4.7221	0.0000

5.1.3 VAR Multistep Forecasts

The dataset `unrate.RData` consists of 684 monthly observations for the period 1968/1–2024/12 for unemployment rate, the dependent variable, and three covariates: IC (Initial Claims for unemployment insurance), inflation rate (seasonal growth

Table 5.2 Diebold–Mariano test for two subperiods: 2008/1–2019/12 and 2020/1–2024/12, for which the loss function is the RMSE. For the pre-COVID period 2008/1–2019/12, ARIMA is the best forecaster and for the post-COVID period, the SVM outperforms the other four methods and the next best model is ARIMA

2008/1–2019/12			2020/1–2024/12		
H_0	Statistic	*p*-value	H_0	Statistic	*p*-value
ENET = **ARIMA**	2.40	0.009	ARIMA = **SVM**	2.60	0.0059
GBM = **ARIMA**	2.72	0.0036	ENET = **SVM**	4.61	0.0000
RF = **ARIMA**	2.57	0.0056	GBM = **SVM**	4.45	0.0000
SVM = **ARIMA**	4.60	0.0000	RF = **SVM**	5.69	0.0000

rate for CPI), and industrial growth (seasonal growth rate for industrial production index). IC is a common indicator for unemployment, and Montgomery et al. (1998) considered it in a multivariate case. Inflation and industrial growth were considered in the work by Edlund and Karlsson (1993) as economic control variables.

The data is received, and the data input is calculated. Examples are the inflation rate and the growth rate of the industrial production index (IPI).

The following code chunk constructs the data input and the timeframe for the rolling window.

Code Chunk. Constructing the data input and the timeframe for the rolling window

```
load("data/unrate.RData")
dat=na.omit(unrate[,1:4])
inf=diff(log(dat[,"CPI"]),12)*100
ICSA=log(dat[,"ICSA"])
ipiG=diff(log(dat[,"IPI"]),12)*100
y=cbind(dat[-seq(12),1],INF=inf, ICSA=ICSA, ipiG=ipiG)
y=na.omit(y)
timeframe=iForecast::rollingWindows(y,
                              estimation="480m",
                              by="6m")
FROM=as.character(timeframe$from)
TO=as.character(timeframe$to)
```

RollingWindows specifies a 480-month period as the initial estimation window and moves 6 months ahead. For this specification, the conventional VAR is estimated using least squares, and this process is termed, LS-VAR.

Econometric VAR: LS and VECM

The rolling structure is the same as that for the previous section. The following code chunk illustrates the process:

Code Chunk. Training using a Rolling LS-VAR

```
library(vars)
FCST0=NULL
for (i in 1:(length(TO)-1)) {
EST=window(y, start=FROM[1], end=TO[i])
VLD=window(y, start=TO[i], end=TO[i+1])[-1,]
out1.var=VAR(EST,lag.max=12,type="none")
fcst_var=predict(out1.var,
                       n.ahead=nrow(VLD))[[1]]$unrate[,1]
fcst_var=as.matrix(fcst_var)
rownames(fcst_var)=rownames(VLD)
fcst_var=as.timeSeries(fcst_var)
FCST0=rbind(FCST0,fcst_var)
}
```

The rolling window is again a loop stack of multistep forecasts using object `FCST0` for each loop. After estimation, the actual values are constructed, and the results are saved.

Finally, in terms of ACF1 $\approx$ 0.8, the unemployment rate may be non-stationary or governed by a unit-root process. A cointegrated VAR can be forecast using a Vector Error Correction Model (VECM). In R, `tsDyn::VECM` is used but Fig. 5.10 shows that this is a very poor forecasting tool of the RMSE $\approx$ 2.99. This shows that a VECM is not a good forecaster for unemployment rate.

Machine Learning VAR

This example uses a SVM to estimate a VAR system using the following code chunk. Line 1 executes `tts.var()`, and Line 2 performs a multistep forecast using `tts.var()`.

Code Chunk. Training using a Rolling ML-VAR

```
FCST0=NULL
for (i in 1:(length(TO)-1)) {
VLD=window(y,start=TO[i],end=TO[i+1])[-1,]
1.  OUT1=tts.var(data=y,
             p=6,
             method="svm",
             train.end=TO[i],
             type="none")
2.  fcst_ml=iForecast.var(OUT1, n.ahead=nrow(VLD))
rownames(fcst_ml)=rownames(VLD)
fcst_ml=as.timeSeries(fcst_ml)
FCST0=rbind(FCST0,fcst_ml)
colnames(FCST0)=c("unrate","inf","ICSA","ipiG")
print(paste0(m, ":",i,"/",(length(TO)-1)))
}
```

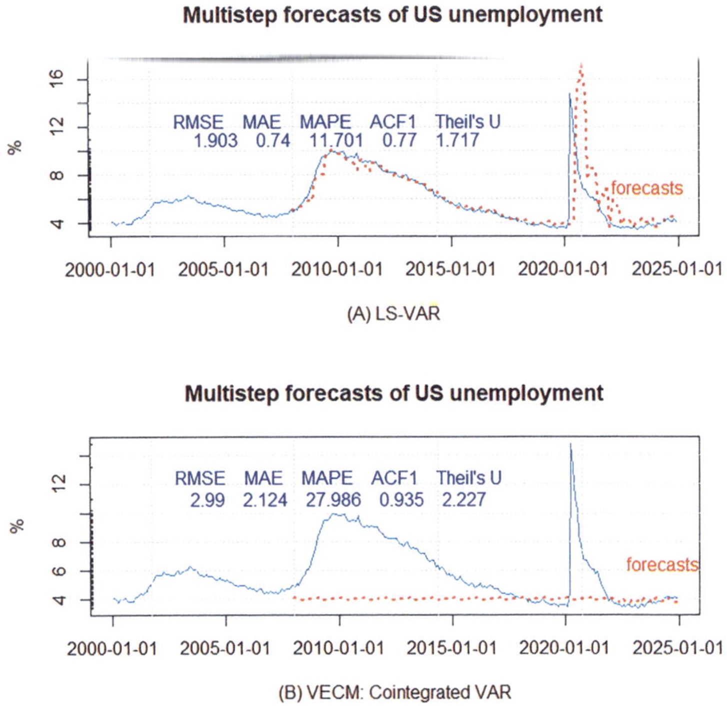

Fig. 5.10 Multistep forecasts using an econometric VAR.: (**a**) A conventional LS-VAR and (**b**) a maximum likelihood estimator for a cointegrated VAR using a VECM. The accuracy measures show that a cointegrated VAR is not a good forecaster and the unemployment rate forecasts are around 4%

	RMSE	MAE	MAPE	ACF1	Theil's U
enet	1.467	0.559	8.299	0.716	1.338
LS	1.903	0.740	11.701	0.770	1.717
svm	1.393	0.641	9.949	0.762	1.384

Fig. 5.11 Comparison of accuracy measures for three VAR forecasters. LS-VAR, a SVM, and ENET are estimated. In terms of the RMSE, a SVM is the best method and in terms of the MAE, ENET outperforms others

A SVM and ENET are also estimated. Figure 5.11 shows the results for several loss functions as accuracy measures. In terms of the RMSE, a SVM is the best method. In terms of the MAE, ENET outperforms others. Interested readers can conduct a Diebold–Mariano test, as advised for Table 5.1.

Figure 5.12 visualizes the forecasting values against the actual values. The results are similar to previous results. The performance for both methods is similar for 2008/1 to 2019/12 and the COVID-19 incident tests robustness. ENET and the SVM perform relatively similar.

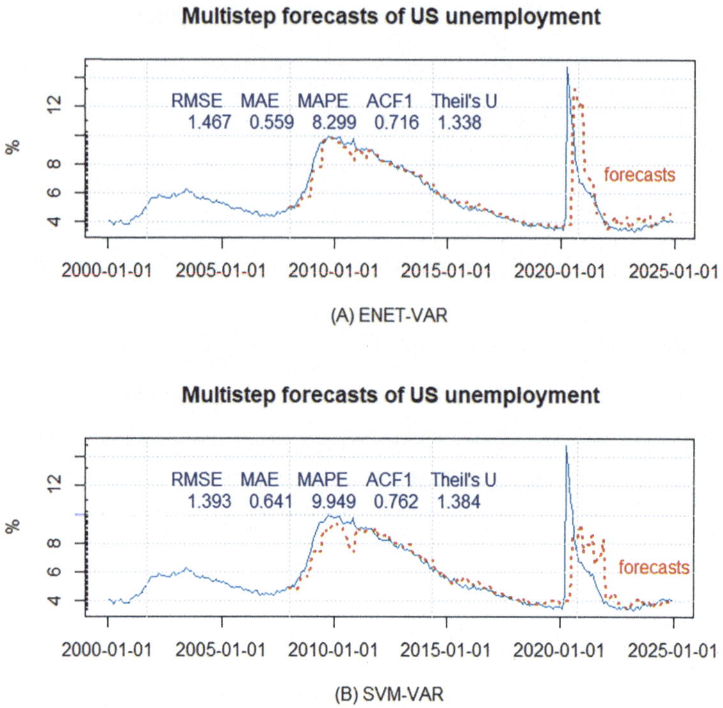

Fig. 5.12 Multistep forecasts using a rolling window VAR and (**a**) ENET and (**b**) SVM. From 2008/1 to 2019/12, both methods perform similarly and the COVID-19 incident tests robustness. ENET and the SVM perform similarly, and Fig. 5.10 shows that LS-VAR is the worst forecaster

5.1.4 VAR Onestep Forecasts

Finally, if it is difficult to forecast the real future because of data availability, onestep ahead prediction may not work for a low-frequency time series so the performance of a onestep ahead prediction is of interest. To implement a onestep ahead prediction, the argument is modified using `b="1m"` within `rollwingWindows`. Figure 5.13 plots the results.

Figure 5.13 shows that with the exception of ACF1, LS-VAR's predictive accuracy measures are worse than those for the SVM-VAR and for periods immediately after COVID, LS-VAR overpredicts the unemployment rate, and the SVM-VAR underpredicts it. After this period, both methods' forecasting values follow the trend. In terms of the ACF1, a machine learning -VAR (0.68) is much better than LS-VAR(0.13). This illustrates a problem with serial correlation for a machine learning VAR.

In summary, for onestep ahead forecasts over a 204-month rolling window, conventional econometric VAR and machine learning VAR perform similarly but the

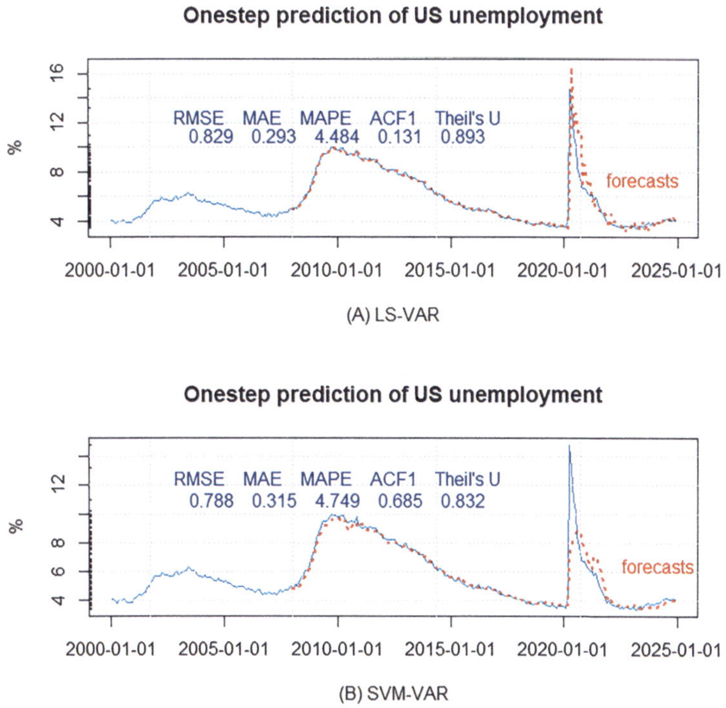

Fig. 5.13 Results for onestep ahead prediction using rolling windows for (**a**) LS- and (**b**) SVM-VAR. LS- and the SVM-VAR perform similarly. Immediately after 2020, the forecasting values for LS-VAR are overpredicted and those for the SVM-VAR are underpredicted. Afterwards, both methods' forecasting values track the trend

multistep forecasts provides better definition and is the better forecaster. Figure 5.12 shows that for the real future, the average of LS and SVM increases the accuracy of the future forecast.

5.2 Predictive Modeling for Index Returns

Financial economics literature notes the prediction of asset returns as problematic because the predictability of returns involves the core issue of financial economics: market efficiency (Fama, 1970). Whether the variables of economic fundamentals can predict the random walk process also challenges theory (Meese & Rogoff, 1983). In practice, the prediction of market returns involves decisions that are related to asset allocation and risk management.

This section determines whether economic activity affects DJ index returns.

5.2.1 Data and Model

Part of the economic variables was measured by Chen et al. (1986). The predictive regression function is specified as:

$$R_t = f(MP_{t-1}, UI_{t-1}, DEI_{t-1}, UPR_{t-1}, UTS_{t-1}, \\ EWIND_{t-1}, VWIND_{t-1}, CG_{t-1}, OG_{t-1}, YP_{t-1}) \tag{5.1}$$

where

R: Excess returns for the DJ index; log returns for DJI minus the risk-free rate from the French data library.[2]
MP: Growth rate for the industrial production index; $MT_t = \ln(IP_t)\text{-}\ln(IP_{t-1})$.
UI: Unexpected inflation; Inf represents the inflation rate, $UI_t = Inf_t - E_{t-1}(Inf_t)$.
DEI: Expected inflation; $DEI_t = E_t(Inf_{t+1}) - E_{t-1}(Inf_t)$.
UPR: Risk premium, defined as the excess return for low-rated corporate bonds (Baa and below) over the 3-month Treasury bill rate; $URP_t = BAA_t - TB3MS_t$.
UTS: Term structure, defined as the yield to maturity for a 10-year bond minus the yield for a 3-month Treasury bill; $UTS_t = GS10_t - TB3MS_t$.
EWIND: The equally weighted industry-wide asset returns, taken from the French data library, as the dependent variable *R*.
VWIND: The value-weighted industry-wide asset returns, taken from the French data, as the dependent variable *R.CG*: Real per capita consumption growth rate.
OG: Growth rate for the crude oil price.
YP: Growth rate for the industrial production index $YP_t = \ln(IP_t) - \ln(IP_{t-12})$.

The descriptive statistics of these variables is listed in Table 5.3, and the time series are plotted in Fig. 5.14. The summary statistics shows that the most volatile variable is the growth rate for the crude oil price te (OG), which has the widest range and the largest standard deviation (Stdev).

The methodological steps are:

Step 1. As a baseline model, index returns are forecast using ARIMA.

Step 2. Index returns are forecast using ARIMAx, which uses Eq. (5.1), similarly to the method of Chen et al. (1986).

Step 3. Three machine algorithms (enet, gbm, and svm) are applied to Eq. (5.2). The model is denoted by ML-ARx. For example, enet-ARx(1) implies AR(1) + X_{t-1}.

$$R_t = f(R_{t-1}, MP_{t-1}, UI_{t-1}, DEI_{t-1}, UPR_{t-1}, UTS_{t-1}, \\ EWIND_{t-1}, VWIND_{t-1}, CG_{t-1}, OG_{t-1}, YP_{t-1}) \tag{5.2}$$

Finally, the forecasting performance for these models is compared.

[2] https://mba.tuck.dartmouth.edu/pages/faculty/ken.french/data_library.html

Table 5.3 Descriptive statistics for the data for 1959/2 to 2024/12. The summary statistics shows that DJI returns is not the most volatile variable: the crude oil price growth rate (OG) has the widest range and largest standard deviation (stdev)

	Mean	Median	Max.	Min.	Stdev	Skewness	Kurtosis
DJI	0.537	0.854	10.377	−23.099	3.56	−1.094	4.759
MP	0.194	0.243	6.377	−14.204	0.996	−3.779	59.449
UI	0.003	−0.017	1.456	−1.67	0.285	−0.194	2.75
DEI	0	−0.002	0.673	−0.637	0.14	−0.061	2.224
UPR	2	1.94	6.01	0.29	0.79	0.832	2.473
UTS	1.415	1.46	4.42	−2.65	1.241	−0.226	−0.362
YP	2.45	2.809	19.612	−18.877	4.903	−0.834	2.381
OG	0.352	0	48.501	−66.894	8.521	−0.939	9.227
CG	0.267	0.3	8.2	−11	0.791	−2.704	74.152
VWRETD	−0.062	0.648	18.737	−26.054	5.779	−0.638	1.345
EWRETD	0.057	0.558	28.349	−28.703	6.56	−0.328	1.629

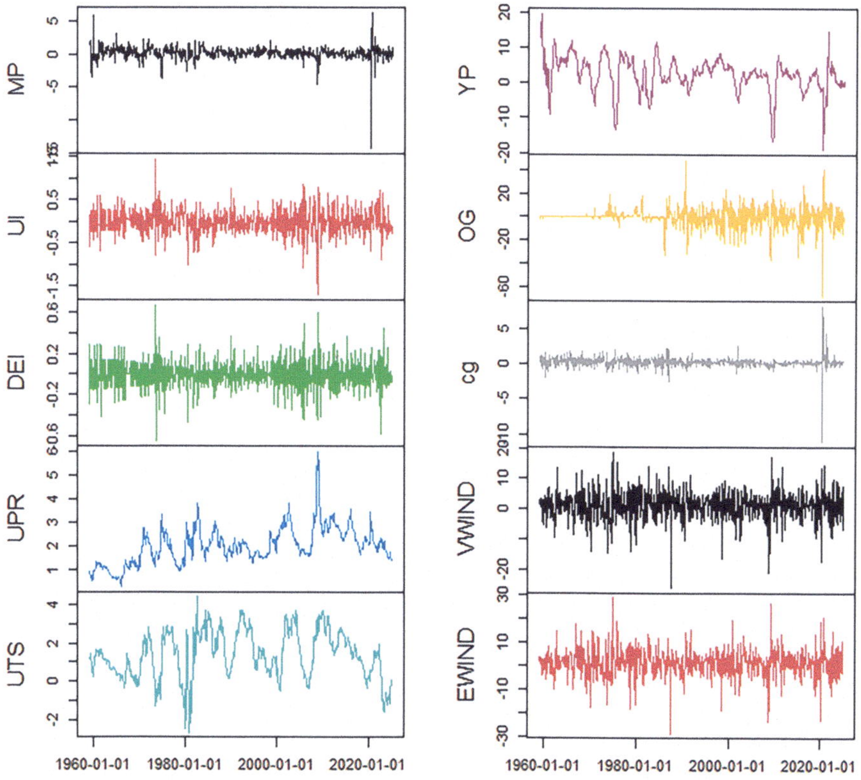

Fig. 5.14 Time series plots for ten covariate variables in Eq. (5.1)

5.2.2 Empirical Outcomes

Univariate Case

As a benchmark model, a pure ARIMA process is fitted by applying `auto.arima` to the DJI returns, and the model is best fitted using MA(1). A onestep forecast is then performed, and the window is moved forward. The results are plotted in Fig. 5.15, where the rolling window is structured as follows: The data is for 1959M2–2024M12, the in-sample training period starts with the first 540 months, and the rolling step is 1 month.

In Fig. 5.15, an additional accuracy measure is added: MAAPE (Mean Absolute Arctan Prediction Error), MAPE usually approaches infinity when the dependent variable is the rate of change because a near-zero percentage value can be produced; so, it usually makes the MAPE too large to be informative. Kim and Kim (2016) proposed MAAPE to remedy this problem and proposed a new measure for forecast accuracy.

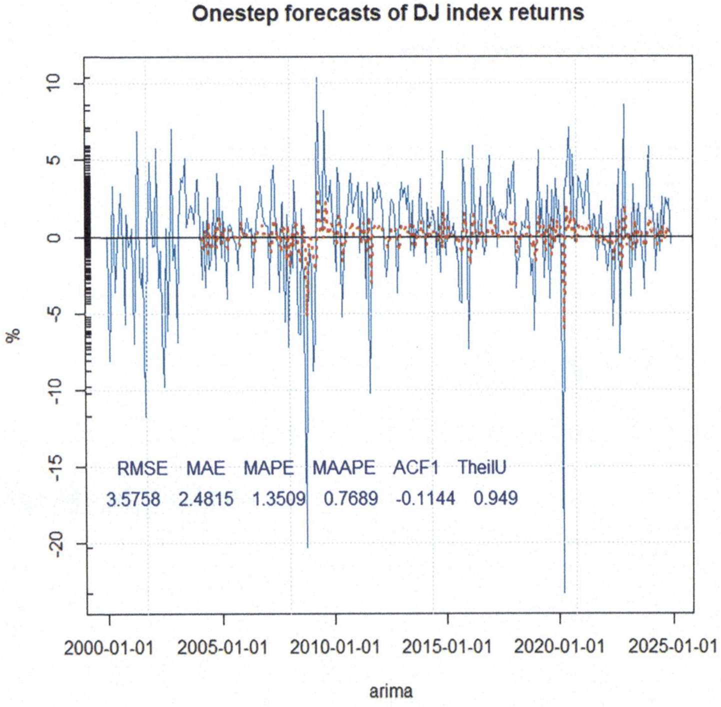

Fig. 5.15 The benchmark model: Onestep forecasts use a pure MA(1). The rolling window is structured as follows: The data is for 1959M2–2024M12, the in-sample training period starts with the first 540 months and and prediction is onestep ahead. The largest negative return is −23.1%, around 2020 when the COVID pandemic began. The red dotted line is the multistep forecast using a pure ARIMA and `auto.arima`. In this case, the MAAPE measure is also reported, which avoids the infinity problem of MAPE

The MAPE is the average of the absolute percentage error (APE). Kim and Kim considered a triangle with adjacent and opposite sides that are, respectively, equal to $|Y|$ and $|E|$, where Y and E are the actual value and the forecast error, respectively. In principle, APE is the gradient of the hypotenuse. Clearly, the slope can be measured either as a ratio of $|E|$ to $|Y|$, which ranges from zero to infinity; or, as an angle from 0 to 90°. The gradient is a ratio, specifically the APE so the angle of the gradient is a useful measure of the forecast accuracy, as Kim and Kim (2016) proposed. For this gradient, the ratio is the tangent of the angle so angle θ can be expressed using $|Y|$ and $|E|$ as:

$$MAAPE = \frac{\sum \tan^{-1}\left|\frac{E}{Y}\right|}{N}$$

where the MAAPE is calculated using the function `Accuracy()` in the package `iForecast`.

As competing models for whether the economic variables contain predictive components, MAx and three machine learning algorithms are used: ENET, a GBM, and a SVM. MAx uses Eq. (5.1), which is a result of `auto.arima` with ten exogenous variables. This is best fitted with ARIMA(0,0,1) plus X so it is denoted as MAx(1). For machine learning methods, ARx(1) is used to denote the specification for machine learning methods, so an AR(1) term and the exogenous variables have a lag order of 1.

Figure 5.16 shows all. A comparison with Fig. 5.15 shows that in terms of the RMSE, MAE, and MAAPE, all competing models with X outperform the MA(1) bench model and secondly, MAx(1) is the best forecaster of the four competing forecasting constituents.

To determine the importance of economic forces, the Diebold–Mariano statistic is used to test for predictive accuracy for the bench model without exogeneous economic variables and competing models with ten exogeneous economic variables. This shows whether a pure MA(1) can be rejected by the other four models.

Table 5.4 shows the results, where the equality null is stated in the same form as in Table 5.1. If RMSE is the loss function, all models reject the benchmark MA(1), and MAx(1) shows the most significant rejection. MAE gives the same results. Table 5.4 shows that using ten exogenous variables improves the predictability of index returns and they unanimously reject the pure ARIMA and coincide with multistep results. Table 5.4 confirms that economic variables contain predictable components of stock returns.

The predictive accuracy of all competing models compared by omitting the benchmark model MA(1) and conducting a pair-wise Diebold–Mariano test on the remaining models. Table 5.5 shows that using both loss functions, of these competing models, MAx(1) outperforms its machine learning counterparts.

Finally, the importance of economic variables is continually evaluated using VAR. The next section continues the study.

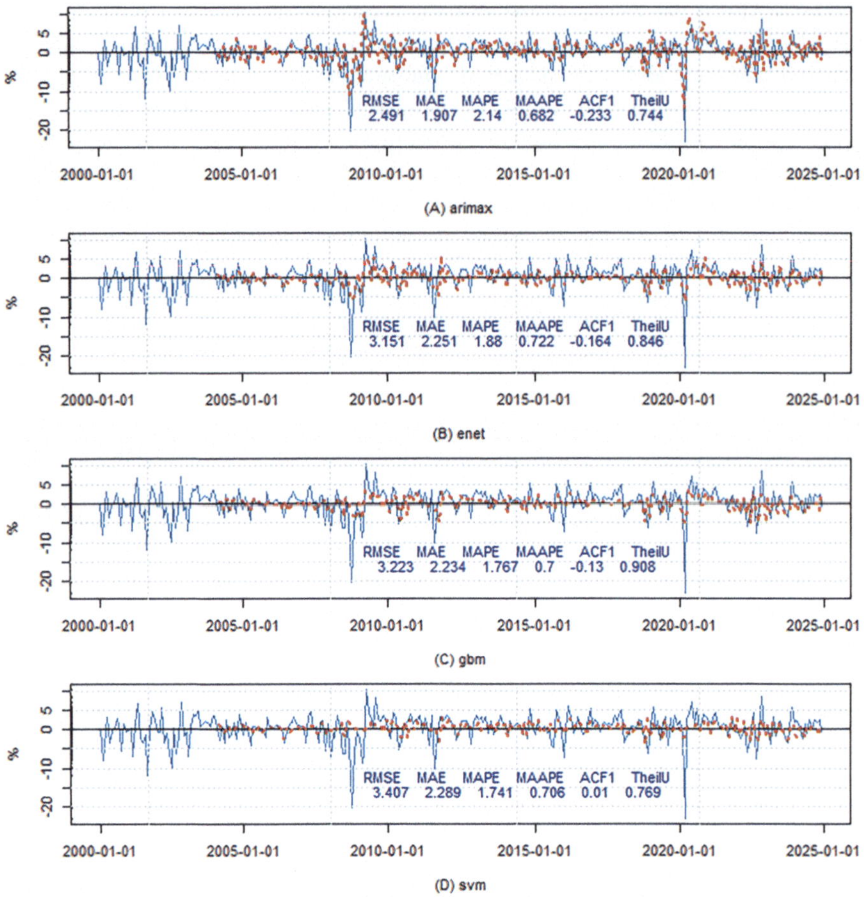

Fig. 5.16 The competing models: Onestep forecasts for DJ index returns: (**a**) MAx(1), (**b**) ARx(1) of elastic net, (**c**) ARx(1) for a gradient boost machine, and (**d**) ARx(1) for a support vector machine. Forecasts are generated using a 3-month ahead forecast for each window. In terms of the RMSE and MAE, MAx(1) is the best forecaster

Table 5.4 The importance of economic forces: a Diebold–Mariano test for onestep ahead forecasts. In terms of the RMSE and MAE, MAx and machine learning methods confirm that economic variables are important in terms of forecasting stock returns

	RMSE		MAE	
H_0	Stat	p-value	Stat	p-value
MA(1) = **MAx**(1)	2.81	0.003	3.87	0.00012
MA(1) = **ENET-ARx**(1)	2.61	0.0048	2.49	0.0068
MA(1) = **GBM-ARx**(1)	2.34	0.01	2.81	0.0027
MA(1) = **SVM-ARx**(1)	1.59	0.057	2.35	0.01

Table 5.5 Onestep ARx forecast evaluation: Diebold–Mariano test of MAx against each machine learning method. For both loss functions, the equality null is significantly rejected by MAx, which shows that the forecasting performance of MAx is superior to that for the machine learning methods

	RMSE		MAE	
H_0	Stat	p-value	Stat	p-value
ENET-ARx(1) = **MAx**(1)	2.36	0.0095	2.51	0.006
GBM-ARx(1) = **MAx**(1)	2.50	0.0066	2.32	0.011
SVM-ARx(1) = **MAx**(1)	2.43	0.0078	2.55	0.006

Multivariate Case of VAR

The multistep results are analyzed using VAR. Figure 5.17 shows that the standard LS-VAR scores the highest in terms of RMSE, MAE, and MAAPE and that all three models have a similar ACF1 value. Their differences in terms of RMSE and MAE are significant.

Using MA(1) as the benchmark model, the Diebold–Mariano statistic is calculated for a pair-wise comparison, Table 5.6 tabulates the results. The forecasting performance for a conventional LS-VAR is significantly less than that for the benchmark model MA(1), which significantly rejects it under the null hypothesis.

Equality null the MA(1) = ML-VAR(6) cannot be rejected using the RMSE, but it can be partly rejected using the MAE. Multistep forecasts give ambiguous results.

In terms of the forecasting performance for the onestep ahead VAR, Fig. 5.18 and Table 5.7 exhibit similar results to those for the multistep. Figure 5.18 shows that the standard LS-VAR's accuracy measures have the greatest values. The differences in terms of RMSE and MAE between LS-VAR and the other two models are also significant.

The results for a pair-wise Diebold–Mariano test are shown in Table 5.7. The forecasting performance is similar for a conventional LS-VAR and MA(1). ML-VAR(6) outperforms MA(1), which shows that the economic forces contain important predictive components of stock returns.

In summary, machine learning extracts significant predictable information from ten covariates, whether it is specified as ML-ARx or ML-VAR. Onestep ahead prediction also gives a better forecasting horizon.

5.3 Volatility Forecasting

This section is an R Exercise section. The data compilation for daily financial time series is presented instead of practice exercises. In general, a financial time series is mainly characterized by volatility. Frequency applies to daily and intra-day transaction data. For finance practitioners, a good forecast is a prerequisite for profitable trades.

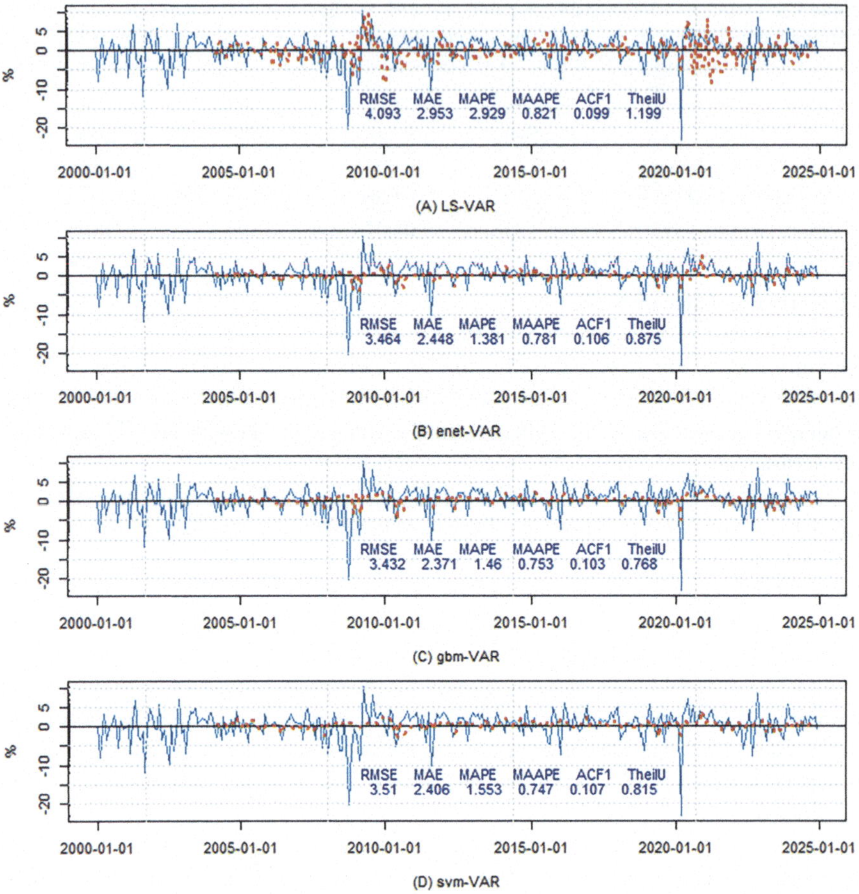

Fig. 5.17 Multistep VAR forecasts for DJ index returns: (**a**) LS-VAR, (**b**) elastic net, (**c**) gradient boost machine, and (**d**) support vector machine. The standard LS-VAR has the largest RMSE (4.093), MAE(2.953), and MAAPE (0.821) values. All three models exhibit a similar ACF1 value. The differences in terms of RMSE and MAE between LS-VAR and the other two models are significant. In terms of Table 5.6, the pair-wise Diebold–Mariano test shows that the forecasting performance for LS-VAR is inferior to that for a svm-VAR or enet-VAR, so machine learning methods are effective

Table 5.6 Evaluation of multistep VAR forecasts using a Diebold–Mariano test against LS-VAR. For both loss functions, the equality null is substantially rejected, so all three machine learning models are superior to LS-VAR

	RMSE		MAE	
H_0	Stat	p-value	Stat	p-value
LS-VAR(6) = **MA**(1)	2.624	0.001	3.15	0.0001
MA(1) = **ENET-VAR**(6)	1.24	0.17	1.37	0.08
MA(1) = **GBM-VAR**(6)	1.15	0.13	2.41	0.008
MA(1) = **SVM-VAR**(6)	0.789	0.215	1.96	0.025

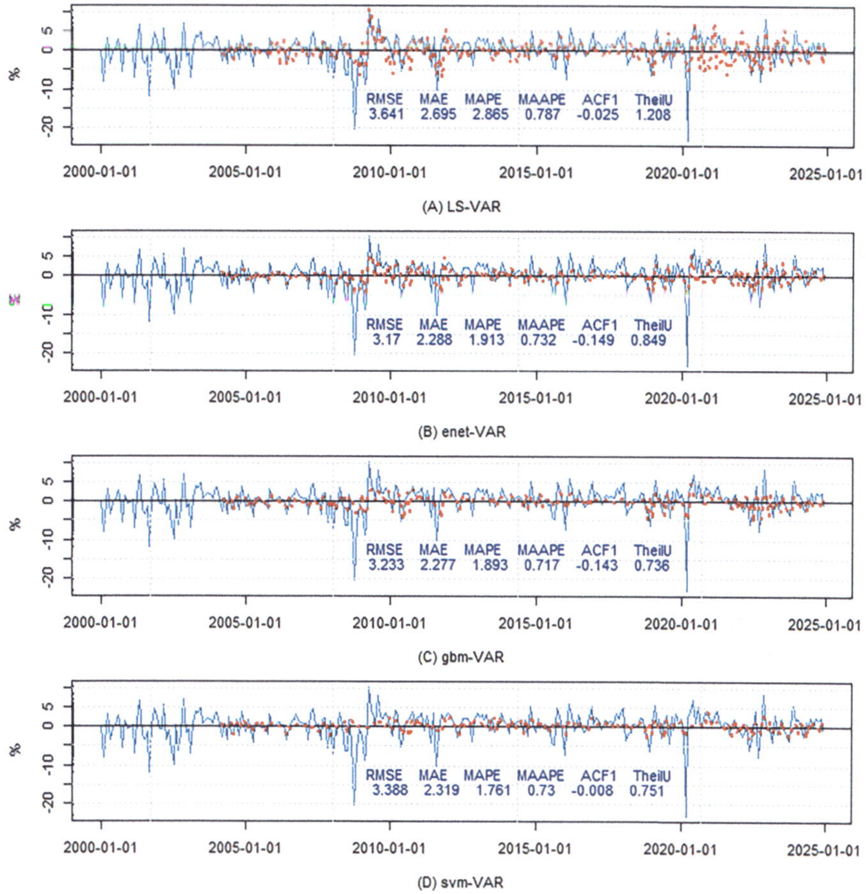

Fig. 5.18 Onestep VAR forecasts for DJ index returns: (**a**) LS-VAR, (**b**) elastic net, (**c**) gradient boost machine, and (**d**) support vector machine. The standard LS-VAR has the largest RMSE, MAE, and MAAPE values. The differences in terms of RMSE and MAE between LS-VAR and the other two models are significant. In terms of Table 5.7, the pair-wise Diebold–Mariano test shows that the forecasting performances of the svm-VAR and enet-VAR are superior to that of LS-VAR, so machine learning methods are effective

Table 5.7 Evaluation of onestep ahead VAR forecasts using a Diebold–Mariano test against LS-VAR. For both loss functions, the equality null is systematically rejected, which shows that machine learning methods best extract the predictive components of economic forces

	RMSE		MAE	
H_0	Stat	p-value	Stat	p-value
LS-VAR(6) = **MA**(1)	0.315	0.48	1.458	0.07
MA(1) = **ENET-VAR**(6)	2.58	0.005	2.146	0.016
MA(1) = **GBM-VAR**(6)	2.38	0.009	2.43	0.0079
MA(1) = **SVM-VAR**(6)	1.909	0.0287	2.102	0.018

time	Ret	MOM	logV	RV	TR	ATR
1980-01-22	-0.5277	5.75	17.7399	0.702	2.18	2.3379
1980-01-23	1.716	8.22	17.742	0.9625	3	2.3852
1980-01-24	0.2289	7.18	17.8942	0.7333	2.32	2.3805
1980-01-25	-0.0792	6.8	17.6678	0.6647	2.09	2.3598
1980-01-28	1.0855	5.9	17.7974	0.8584	2.72	2.3855
1980-01-29	-0.6815	5.02	17.8315	0.8639	2.74	2.4108
...	...	...	...	...	...	...
2025-03-28	-1.9935	-33.6201	22.2303	0.7271	120.8901	90.8869
2025-03-31	0.5523	39.7803	22.3829	0.9009	138.8301	94.3114
2025-04-01	0.3774	33.77	22.2127	0.5924	92.0498	94.1499
2025-04-02	0.6706	149.4502	22.1687	0.7928	123.8301	96.2699
2025-04-03	-4.9606	-242.4199	22.6988	0.72	280.1401	109.4035
2025-04-04	-6.1609	-601.04	22.9041	1.5474	326.6201	124.919

Fig. 5.19 S&P500 dataset, 1980/1/2 to 2025/4/4, where Ret denotes the log returns for the S&P500 closing price, MOM is the momentum for 14 days, LogV is log(Volume), RV is the Parkinson's range-based volatility, TR is the true range, and ATR measures the average true range for 14 days

The dataset uses the daily S&P 500 from 1980/1/2 to 2025/4/4: Open, High, Low, Close, Volume, and Adjusted Close. This dataset contains more than 10,000 observations, and the compiled data table is shown in Fig. 5.19. The last 2 days are the dates on which tariffs were set by US President Trump.

Definitions of variables in Fig. 5.19 are:

Ret: the log returns for the S&P500 closing price
MOM: momentum: 14 days
LogV: log(Volume)
RV: Range-based volatility for Parkinson (1980)
TR: True range
ATR: Average true range for 14 days

This dataset compiles some popular measures for stock time series. RV, TR, and ATR are volatility measures. Figure 5.20 plots a visual presentation, which shows that over time, each series exhibits different dynamic patterns, such as unexpected jumps, irregular trends, and clusters. These out-of-sample characteristics complicate forecasting.

The R Code for this dataset is listed below. Interested readers are encouraged to build models and calculate forecasts. Volatility forecasting is a time series that is difficult to predict, and machine learning methods do not show any suitability for volatility forecasting so interested readers are encouraged to begin with the return series, RV, and ATR.

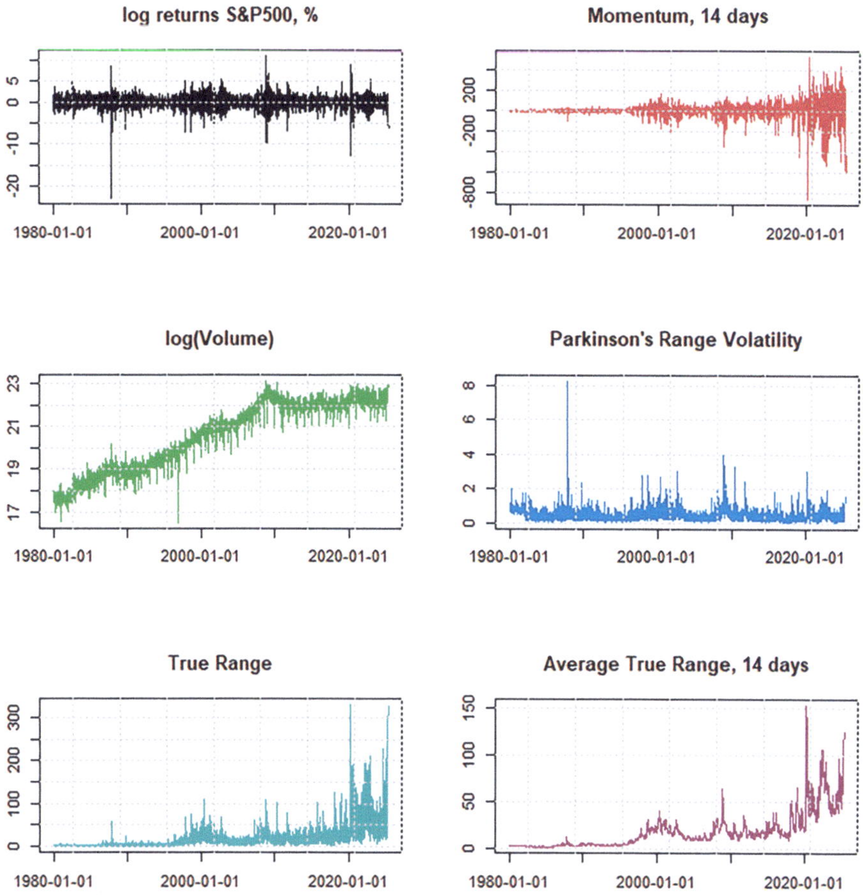

Fig. 5.20 Time series plots for the S&P500, 1980/1/2 to 2025/4/4. In terms of several volatility measures over time, underlying dynamic patterns drive these series, such as unexpected jumps, irregular trends, and clusters. Ret denotes the log returns for the S&P500 closing price, MOM is the momentum for 14 days, LogV is log(Volume), RV is the range-based volatility for Parkinson (1980), TR is the true range, and ATR is the average true range for 14 days

```
library(timeSeries)
library(xts)
quantmod::getSymbols("^GSPC", from="1980-01-01")
GSPC=as.timeSeries(GSPC)

Ret=TTR::ROC(GSPC[,6], n=1,na.pad = TRUE)*100
mom=TTR::momentum(GSPC[,6], n = 14, na.pad = TRUE)
Ret=as.timeSeries(Ret)
mom=as.timeSeries(mom)
logV=log(GSPC[,5]) #Log Volume
```

```
rownames(Ret)=rownames(mom)=rownames(logV)=rownames(GSPC)
RV=((log(GSPC[,2]/GSPC[,3]))/(4*log(2)))*100
ATR=as.timeSeries(TTR::ATR(GSPC[,c(2:4)], n=14))

SP500_daily=na.omit(cbind(Ret=Ret,
                    MOM=mom,
                    logV=logV,
                    RV=RV,
                    TR=ATR[,"tr"],
                    ATR=ATR[,"atr"]))
head(SP500_daily)
```

References

Aiolfi, M., Carlos, C., & Timmermann, A. (2011). Forecast combinations. In M. Clements & D. Hendry (Eds.), *The Oxford handbook of economic forecasting* (pp. 355–388). Oxford university Press.

Bates, J. M., & Granger, C. W. J. (1969). The combination of forecasts. *Operational Research Quarterly, 26*, 325–329.

Bergmeir, C., Hyndman, R. J., & Benítez, J. M. (2016). Bagging exponential smoothing methods using STL decomposition and Box–Cox transformation. *International Journal of Forecasting, 32*(2), 303–312.

Bishop, C. M. (2006). *Pattern recognition and machine learning*. Springer.

Boehmke B. and Greenwell B. (2020) Hands-on Machine Learning with R. Chapman & Hall/ CRC Press, FL.

Bontempi, G., Birattari, M., & Bersini, H. (1999). Local learning for iterated time-series prediction. In I. Bratko & S. Dzeroski (Eds.), *Machine learning: Proceedings of the sixteenth international conference* (pp. 32–38). Morgan Kaufmann Publishers.

Box, G. and Jenkins, G. (1970) Time Series Analysis: Forecasting and Control. Holden-Day, San Francisco.

Chen, N.-F., Roll, R., & Ross, S. A. (1986). Economic forces and the stock market. *Journal of Business, 59*(3), 383–403.

Chevillon, G. (2007). Direct multi-step estimation and forecasting. *Journal of Economic Surveys, 21*(4), 746–785.

Davis, R. A., Holan, S. H., Lund, R., & Ravishanker, N. (Eds.). (2016). *Handbook of discrete-valued time series* (1st ed.). Chapman and Hall/CRC. https://doi.org/10.1201/b19485

De Livera, A. M., Hyndman, R. J., & Snyder, R. D. (2011). Forecasting time series with complex seasonal patterns using exponential smoothing. *Journal of the American Statistical Association, 106*(496), 1513–1527.

DeMiguel, V., Garlappi, L., & Uppal, R. (2009). Optimal versus naive diversification: How inefficient is the 1/N portfolio strategy? *Review of Financial Studies, 22*, 1915–1953.

Diebold, F. X., & Mariano, R. S. (1995). Comparing Predictive Accuracy. *Journal of Business and Economic Statistics*, 13, 253–263.

Edlund, P.-O., & Karlsson, S. (1993). Forecasting the Swedish unemployment rate: VAR us. transfer function modelling. *International Journal of Forecasting, 9*(1), 61–76.

Fama, E. F. (1970). Efficient capital markets: A review of theory and empirical work. *Journal of Finance, 25*(2), 383–417.

Franses, P. H., & Van Dijk, D. (2000). *Non-linear time series models in empirical finance.* Cambridge University Press.

T.-w. Ho, *Time Series Forecasting using Machine Learning*,
https://doi.org/10.1007/978-3-031-97946-0

Friedman, J. H. (2001). Greedy function approximation: A gradient boosting machine. *Annals of Statistics, 29*, 1189–1232.

Hansen, B. E. (2007). Least squares model averaging. *Econometrica, 75*(4), 1175–1189.

Hansen, B. E., & Racine, J. S. (2012). Jackknife model averaging. *Journal of Econometrics, 167*(1), 38–46.

Haslett, J., & Raftery, A. E. (1989). Space-time modelling with long-memory dependence: Assessing Ireland's wind power resource (with discussion). *Applied Statistics, 38*, 1–50.

Hastie, T., & Tibshirani, R. (1990). *Generalized additive models*. Chapman and Hall.

Hastie, T., Tibshirani, R., and Friedman, J. (2009). *The Elements of Statistical Learning: Data Mining, Inference, and Prediction*. Spring-Verlag.

Hastie, T., Tibshirani, R., & Friedman, J. (2017). *The elements of statistical learning: Data mining, inference, and prediction*. Springer Science & Business Media.

Hyndman, R. J., & Athanasopoulos, G. (2018). *Forecasting: Principles and practice* (2nd ed.). Retrieved from https://otexts.com/fpp2/.

Hyndman, R. J., & Khandakar, Y. (2008). Automatic time series forecasting: The forecast package for R. *Journal of Statistical Software, 26*(3).

Hyndman, R. J., & Koehler, A. B. (2006). Another look at measures of forecast accuracy. *International Journal of Forecasting, 22*(4), 679–688.

Hyndman, R. J., Athanasopoulos, G., Bergmeir, C., Caceres, G., Chhay, L., O'Hara-Wild, M., & Yasmeen, F. (2018). *Forecast: Forecasting functions for time series and linear models*. Retrieved from https://rdrr.io/cran/forecast/.

Kim, S., & Kim, H. (2016). A new metric of absolute percentage error for intermittent demand forecasts. *International Journal of Forecasting, 32*(3), 669–679.

Kuhn, M., & Johnson, K. (2013). *Applied predictive modeling*. Springer.

Liao, J.-C., & Tsay, W.-J. (2020). Optimal multi-step VAR forecasting averaging. *Econometric Theory, 36*(6), 1099–1126.

Maharaj, E. A., D'Urso, P., & Caiado, J. (2019). *Time series clustering and classification*. Chapman and Hall/CRC.

Makridakis, S. (1993). Accuracy measures: Theoretical and practical concerns. *International Journal of Forecasting, 9*(4), 527–529.

McNames, J. (1998). A nearest trajectory strategy for time series prediction. In *Proceedings of the international workshop on advanced black-box techniques for nonlinear modeling* (pp. 112–128). K.U. Leuven.

Meese, R. A., & Rogoff, K. (1983). Empirical exchange rate models of the seventies: Do they fit out of sample? *Journal of International Economics, 14*(1–2), 3–24.

Montgomery, A. L., Zarnowitz, V., Tsay, R. S., & Tiao, G. C. (1998). Forecasting the U.S. unemployment rate. *Journal of the American Statistical Association, 93*(442), 478–493.

Parkinson, M. (1980). The extreme value method for estimating the variance of the rate of return. *Journal of Business, 53*(1), 61–65.

Pesaran, M. H., & Pick, A. (2011). Forecast combination across estimation windows. *Journal of Business and Economic Statistics, 29*(2), 307–318.

Pflug, G. C., Pichler, A., & Wozabal, D. (2012). The 1/N investment strategy is optimal under high model ambiguity. *Journal of Banking and Finance, 36*, 410–417.

Saad, E., Prokhorov, D., & Wunsch, D. (1998). Comparative study of stock trend prediction using time delay, recurrent and probabilistic neural networks. *IEEE Transactions on Neural Networks, 9*(6), 1456–1470.

Sims, C. A. (1980). Macroeconomics and Reality. *Econometrica, 48*(1), 1–48.

Stock, J. H., & Watson, M. W. (1998). *A comparison of linear and nonlinear univariate models for forecasting macroeconomic time series (No. w6607)*. National Bureau of Economic Research.

Stone Charles J. (1985) Additive Regression and Other Nonparametric Models. *Annals of Statistics, 13*(2), 689–705.

Taylor, J. W. (2003). Short-term electricity demand forecasting using double seasonal exponential smoothing. *Journal of the Operational Research Society, 54*(8), 799–805.

Tibshirani, R. (1996). Regression shrinkage and selection via the lasso. *Journal of the Royal Statistical Society Series B (Methodological), 58*(1), 267–288.

Tsay, Ruey S. (2010) *Analysis of Financial Time Series*. 3rd Edition, John Wiley & Sons, Hoboken.

Tu, J., & Zhou, G. (2011). Markowitz meets Talmud: A combination of sophisticated and naive diversification strategies. *Journal of Financial Economics, 99*(1), 204–215.

Vapnik, V. (2000). *The nature of statistical learning theory* (2nd ed.). Springer-Verlag.

Weiss, C. H. (2018). *An introduction to discrete-valued time series*. Wiley.

Wood, S. N., & Augustin, N. H. (2002). GAMs with integrated model selection using penalized regression splines and applications to environmental modelling. *Ecological Modelling, 157*(2–3), 157–177.